Civil-Military Cooperation and International Collaboration in CYBER OPERATIONS

INSTITUTE FOR LEADERSHIP
AND STRATEGIC STUDIES

Symposium Monograph Series

Published by:
University of North Georgia Press
Dahlonega, Georgia

Printing Support by:
Lightning Source Inc.
La Vergne, Tennessee

Cover and book design by Corey Parson.

ISBN: 978-1-940771-51-9

Printed in the United States of America, 2017

For more information, please visit: http://ung.edu/university-press
Or e-mail: ungpress@ung.edu

Blue Ridge | Cumming | Dahlonega | Gainesville | Oconee

CONTENTS

Introduction

Dr. Billy Wells, COL (Ret.) USA

The Institute for Leadership and Strategic Studies (ILSS) has a mission to promote research and dialogue related to the important security issues of our day. This involves both undergraduate research as well as graduate programs and international partnerships, and is an essential element of our educational mission as a senior military college. Each year, ILSS hosts a symposium focused on a critical issue, bringing not only scholars in the field together, but also future military officers from nearly a dozen foreign countries. While not included in this symposium monograph, undergraduate participants, particularly cadets and midshipmen, are encouraged to provide poster presentations for the symposium. Each year, the theme looks at a different issue but one related to civil-military cooperation and the thornier issues facing the future leadership of our Nation and the Armed Forces as well as our allies.

The Cyber Domain, this year's focus, is a relatively new field of conflict. While it is in many ways an extension of electronic warfare and signals intelligence, it is also dramatically different. This domain is not limited by geography and can impact operations anywhere on the globe (and in space) at the speed of light. It is also a major "equalizer" among nations. Relatively small countries with far fewer military, economic, or geographic elements of national power can be highly competitive in this domain, and several already are. In fact, preeminence in this domain, which does not rely on geography, can negate both military and, especially, economic power and can directly or indirectly influence political power through various means, including but not limited to social media.

This year's theme, "Civil-Military Cooperation and International Collaboration in Cyber Operations," highlights the difficulties our nation, indeed, the global community, face in dealing with cyber challenges to our security, including the economic, military, and political implications. Equally, the symposium highlights the necessity across all government agencies and the public sector for coordination in an area that requires employing all national and international capabilities in a synchronized effort to protect society and, as necessary, to attack aggressors in the cyber domain. While the military has exceptional capabilities in this area, they are far outpaced by the volume and talent operating in the civilian domain. Hence, there is an extraordinary leadership challenge associated with cyber operations.

It is hopeful that this symposium monograph will be of interest and use to practitioners and researchers involved in this new and exciting domain. We look forward to feedback on this issue as well as to the next symposium, which will again deal with an exceptionally challenging topic facing our military and civilian leaders.

Dr. Billy Wells, COL (Ret.) USA
Oct 21, 2018
Dahlonega, GA

1

Deterrence in the Era of Cyber Weaponization

Dr. Bryson R. Payne, University of North Georgia
Dr. Edward L. Mienie, University of North Georgia
Victor C. Parker, Jr., University of North Georgia

Abstract

The Cold War was in part maintained by the clear nuclear deterrence of assured mutual destruction among superpowers and fear of nuclear reprisal among non-nuclear states, but the era of cyber weaponization is markedly different from the Cold War in a number of ways. Chief among these is that there is no longer as clear a deterrent in cyberwarfare as there was during the nuclear age. There are neither stockpiles of highly-visible, highly-effective cyber weapons that ensure a nation's security by their destructive capacity, nor guaranteed repercussions from striking a similarly-armed or stronger peer. In this paper the authors focus on cyber actions of nation-states as they support the argument that cyberspace cannot be viewed separately from the geopolitical world within the context of the challenges that cyber poses to just war theory and deterrence theory. The authors propose that government and private sector collaboration is crucial to establishing and maintaining both cyber deterrence and cyber defense, with an emphasis on training qualified civilians in cyber tactics, thereby sustaining a corps of well-trained cyber guardians to protect critical systems at home and support both defensive and offensive operations against foreign adversaries.

Introduction

Nuclear deterrence during the Cold War was primarily based on the threat of immediate, assured, and mutual annihilation by a similarly-armed or superior adversary. Deterrence in cyberwarfare in the twenty-first century differs from nuclear deterrence during last century's Cold War in several ways. First, unlike nuclear warheads, cyber weapons may become obsolete before they are ever fired, as computer systems are patched and updated to correct vulnerabilities. Second, an entire class of cyber weapons can become useless after a single deployment, as the executable code can be reverse-engineered and its exploits defended against (or modified and reused by the intended victim). Third, cyber weapons can be employed with less political risk and potentially without human casualties. Fourth, cyber weapons can be developed more rapidly and less expensively than warheads, across borders often with little chance of detection. Finally, there are neither stockpiles of highly-visible, highly-effective cyber weapons that ensure a nation's security by their destructive capacity, nor guaranteed repercussions from striking a similarly-armed peer, or even a superior adversary.

Further, in the absence of an enemy claiming responsibility, attribution of cyber-attacks is a slow and difficult process. Attribution is especially difficult in cyber because there is little physical evidence (electronic dust) and it may take days, weeks, or months after the attack is even detected. Even when we can accurately identify the attacker, Just War Theory (JWT) does not yet consider most non-kinetic cyber-attacks as acts of war permitting proportionate retaliation. More recently, Russia has allegedly attacked critical infrastructure in the Ukraine and other states using cyber weapons, with little to no publicly-disclosed retaliation from victim states or from the international community. Less-developed nations, terrorist groups, and non-state actors can disproportionately employ cyber weapons against a more developed, highly cyber-dependent adversary with little fear of retaliation. For example, if North Korean attackers were able to shut down a US stock exchange, what would be a proportionate response, as North Korea is among twenty or so nations with no stock exchange of their own? A cyber-attack on critical infrastructure in the US or another major world power could be far more devastating than in a less-developed nation. In addition, non-state actors' involvement in cyberwar poses a particular challenge to existing international law because of the unconventional nature of this new form of warfare.

The focus of this paper is on the cyber actions of nation-states in support of the argument that cyberspace cannot be viewed separately from the geopolitical world (Goodman, 2010). An attacker's tactical and strategic goals have to be taken into consideration. We first examine the problems surrounding deterrence in cyberwarfare by reviewing deterrence theory and dissecting the particular difficulties introduced by cyber weaponization. Second, we look in depth at Just War Theory and both how cyber fits into JWT and how JWT must be updated to reflect the complex realities of cyber espionage, cybercrime, and cyberwarfare. Finally, we discuss approaches to resolving some of the gaps in cyber deterrence, including non-cyber interventions, with an emphasis on producing a trained, cyber-competent populace both for national competitive advantage and for national security.

Background

As we grapple with the question of whether or not cyber-attacks should be considered an act of aggression as expected within the traditional parameters of what constitutes an act of war, we face the uncertainty of not knowing precisely who initiated the attack. An added layer of complexity around a cyber-attack is determining whether or not the attacker is a hostile state actor or a rogue criminal actor and determining the initial intent of the attacker. Is it for monetary gains or the furthering of ideological aims or a mischievous event? And to further complicate the challenge, is the ultimate intent of the attacker to cause physical harm or to disrupt the economy, or did they even consider these potential consequences? These are complex challenges that require an informed response.

Just causes of war should have moral limits defining when and how wars ought to be fought (May & Delston, 2016, p. 186). Just war theory dictates that two moral judgments be made when an act of aggression is encountered from a foreign actor. First, is a nation justified in going to war – also called *jus ad bellum* (justice of war)? Does the war satisfy the traditional conditions for declaring war, namely, "(1) just cause, (2) proportionality, and (3) last resort" (May & Delston, 2016, p. 190). A critical element that has to be present is a current act of military aggression from a foreign actor toward another against which a defensive war must be waged for protection against the aggressor.

Second, is a nation fighting the war in a just manner – *jus in bello* (justice in war)? A war that is justified fighting must be fought in a just way (May & Delston, 2016). We see that just war theory focuses on the use of military force and the morality thereof and by default cannot be applied to a cyber-attack where there is no use of military force. This raises the questions: (1) when is a cyber-attack considered an act of war against which at least defensive actions may be taken, and (2) how can these cyber-attacks be deterred? While we struggle to contain this new frontier of warfare and consider where on the spectrum of what constitutes an act of aggression cyber-attacks fall, we need to be intentional about shaping our deterrence to such potential attacks.

Deterrence Theory

Deterrence theory dates back to Thomas Hobbes (1600's), Cesare Beccaria (1700's), and Jeremy Bentham (1800's), all three of whom posit that the effectiveness of deterrence depends upon the severity of punishment/retaliation, the certainty of punishment, and the celerity of punishment. In conventional warfare, force strength in terms of manpower and firing power serve as a deterrence to military aggression by a foreign actor. The threat of the use of force could serve as a deterrence, as it did during the nuclear age, and the threat of mutual annihilation was real and transparent. But what about the recent dawn of cyber-attacks? Is there deterrence to speak of? And to what level do we deter by denial (Snyder, 1961), discouraging cyber-attacks because our enemies know they will not be successful?

Deterrence theory in the nuclear age steered the US and the Soviet Union away from winning wars and towards preventing wars. Today, a cyber-attack poses a threat to "a wide variety of political and military ends" (Geers, 2010, p. 298). A cyber-attack is defined "as any unauthorized access that causes a system to be 'disrupted, degraded, denied or destroyed'" (National Catholic Reporter, 2017). However, the question should be posed, when is a cyber-attack weaponized that would require an appropriate response from the actor being attacked (National Catholic Reporter, 2017)? International law has to mature in addressing this new frontier of warfare by creating parameters within which nation-states are able to observe and maintain the world order as is the case of the individual confronted by physical aggression within a legal framework for maintaining the social order. Domestic criminal laws need to be improved in the quest to hold perpetrators of cyber-attacks responsible. Geers (2010) posits that "denial and punishment" serve as "two primary deterrence strategies" each with three basic requirements: "1. Capability, 2. Communication, and 3. Credibility" (p. 299).

Cyber-attack tools are much easier to acquire, hide, and deploy than nuclear weapons. The testing of a nuclear bomb is almost impossible to hide, whereas a cyber-attack can be tested in a laboratory, with anonymity, and is not bound to any one specific geographical area or any specific time. The Treaty on the

Non-Proliferation of Nuclear Weapons (NPT) regime and the International Atomic Energy Agency (IAEA) play a major role in anti-proliferation of nuclear weapons. However, the challenge to counter cyber-attack tools' proliferation is the question of defining malicious code (Greer, 2010). The denial capability is far more complex to thwart cyber-attacks than is the case with nuclear weapons. We should keep in mind that cyber-attacks may fall below the threshold considered necessary for military retaliation, such as the denial of service attacks, economic espionage, election tampering, and disinformation, to mention a few.

Denial through communication is the second requirement that should be considered in our grand strategy to thwart cyber-attacks (Geers, 2010). The main objective of the Council of Europe Convention on Cybercrime is to "pursue a common criminal policy aimed at the protection of society against cybercrime, especially via national legislation and international cooperation" (Geer, 2010, p. 300). For the Convention to be successful, liability of signatory states cannot extend beyond their national borders and we should keep in mind that hacker software tools do qualify for dual-use purposes (Geer, 2010).

Denial through credibility is the third requirement that should be considered in our grand strategy to thwart cyber-attacks (Geers, 2010). Here the nation-state that is being threatened "must believe that the threat of retaliation – or of a preemptive strike – is real" (Geers, 2010, p. 300). This deterrence requirement is the most complex to assess as it involves international political-military affairs, human psychology, likelihood of miscalculation, and rationality (Geers, 2010).

Goodman (2010) posits that deterrence has eight elements: "an interest, a deterrent declaration, denial measures, penalty measures, credibility, reassurance, fear, and a cost-benefit calculation" (p.105). The deterrent declaration consists of a clear understanding by adversaries that there will be specific consequences to their cyber-attack in the form of denial or penalty measures, or a combination of both (Goodman, 2010). The declaration will only be taken seriously if it is credible and reassuring in the sense that the nation-state does not have to fear retaliation if it refrains from attacking the interests of the other nation-state (Goodman, 2010). Denial and penalty measures should be clearly understood by the adversary so that the actor would be less inclined to engage in a cyber-attack. Moreover, the adversary should also weigh up the costs and benefits of restraint versus action (Goodman, 2010).

Punishment as a strategy is one of last resort as by implication deterrence by denial failed. This simply implies that by threatening greater aggression the initial aggression is prevented. The complication in this strategy is that punishment by attribution is very difficult to achieve because of the anonymity with which cyber-attacks could be launched. This undermines a state's capability to respond to the attack as the victim must know for sure who launched the attack. Convicting innocent parties in a cyber-attack should be avoided at all costs (Goodman, 2010). Punishment is the offensive capacity of deterrence and is made up of interdependency, retaliation, and counter productivity (Goodman, 2010). The attacker needs to weigh up the costs and benefits of a counterattack. Both the attacker and attacked need to determine beforehand how their tactical and strategic goals are served by launching and defending against an attack. Tactical and strategic goals may become incompatible with each other within a geopolitical context. According to classical deterrence theory, punishment needs to be certain, immediate, and severe to serve as a deterrent (Goodman, 2010). However, unlike a nuclear attack, a cyber-attacked state will still have the capacity to retaliate and retaliation does not necessarily mean an overwhelming, disproportionate attack and therefore only certainty of punishment need be established. Individuals, terrorist cells, and criminal organizations can all access cyber weapons, not just nation states.

Punishment by capability is hamstrung when states do not or cannot cooperate. International cooperation is required to be able to assist in cyber investigations (Goodman, 2010). Honoring international agreements becomes a vital part in assigning justice to the perpetrators of cyber-attacks. When a nation-state reneges on an international cooperation agreement and perpetrators of a cyber-attack escape justice because of the non-compliance to the agreement, that nation-state has to assume responsibility for the cyber-attack by implication (Goodman, 2010). We see that without international and domestic law backing, cyber deterrence cannot be effective. We may ask what constitutes a cyber-attack and come to understand that it is any unauthorized access to or through computer networks that causes a system to be disrupted, degraded, denied, or destroyed.

Just War Theory: Challenges Posed by Cyber

Just war theory is based on the premise that while some wars are morally defensible, others are not (Crisher, 2005, p. 2). As B. Crisher writes, "For example, a war of aggression is seen as unjust, while a war to liberate a people from occupation is seen as just. Hence, [just war theory] is a normative theory that distinguishes between just and unjust on moral grounds" (Crisher, 2005, p. 2). Just war theory has as its two main components the principles of *jus ad bellum* (the right to go to war) and *jus in bello* (right conduct within a war) (Crisher, 2005, p. 2). Citing Michael Walzer's cornerstone work in the field (Walzer, 1977), Major Richard P. DiMeglio states that *jus ad bellum* requires us to make judgments about aggression and self-defense; *jus in bello*, about the observance of customary and acceptable rules of engagement (DiMeglio, 2005). Therefore, just war theory is based on moral and ethical principles that govern the reasons why nations can go to war, and the actions conducted by a nation or group of nations while at war.

The first component of just war theory, *jus ad bellum*, lays out several conditions under which nations may resort to war. First is proper authority and public declaration. This involves the idea that war should be declared by an actor with responsibility for public order, such as a national government (Crisher, 2005). Proper authority is complicated by the lack of established international norms regarding cyber issues, as well as the absence of physical borders in the virtual world of cyber networks. Second, the right to go to war should be based on the right intention and just cause (Williams, Jr. & Caldwell, 2006). Just cause involves the idea that a nation must have a morally acceptable justification for going to war. Has the nation itself been attacked? Does the nation need to go to war to protect its citizens? Just cause generally deals with kinetic attacks (bombs, etc.). What threshold level is required when dealing with issues such as cyber theft, critical infrastructure disruption, or, perhaps, election tampering? Does inconveniencing or hurting millions reach the level of moral harm of killing thousands? International legal scholars have not yet reached a consensus on this issue. The answer will almost certainly depend on the facts of each case, e.g., traffic signals being manipulated versus thousands of pacemakers being negatively affected versus disruption of service for one day for Amazon or Walmart in the US, Alibaba in China, or Magnit in Russia.

A third principle under *jus ad bellum* is the probability of success. The purpose of this principle is "to prevent irrational resort to force or hopeless resistance when the outcome of either will clearly be disproportionate or futile" (Crisher, 2005, p. 7). A nation must have a fair probability of success during the engagement, or it should not resort to war. If the citizens of a nation will be crushed due to the superiority of the opponent, or if a large nation retaliates disproportionately against a smaller aggressor, then resorting to war may be deemed immoral. Fourth is the principle of proportionality: "[T]he proportionality principle requires that the probable good consequences achieved by war should outweigh the probable harmful consequences caused by it" (Lango, 2005, p. 263). This concept is similar to certain utilitarian ideas involving the greatest good for the greatest number of people. A fifth principle of *jus ad bellum* is the concept of last resort. Before going to war, have other reasonable, nonviolent options been considered, applied, and exhausted? But what is meant by the term "reasonable" here? There should at a minimum be a reasonable expectation of success: "[W]hen we reasonably expect that a measure will succeed, we also have to recognize that there is a significant risk that it will fail" (Lango, 2005, p. 261). Before resorting to war, the standard of sufficiency is that "there is no reasonable expectation that additional non-military measures will be successful" (Lango, 2005, p. 261). This principle may work in favor of cyber retaliation as this may be seen as a use of force that does not rise to the level of an all-out kinetic war.

How should *jus ad bellum* be applied in the context of cyber operations? According to M. N. Schmitt, "In 2009, the NATO Cooperative Cyber Defence Centre of Excellence (NATO CCD COE), a renowned research and training institution based in Tallinn, Estonia, invited an independent group of experts to produce a manual on the international law governing cyberwarfare. . . . [T]he effort resulted in the publication of the Tallinn Manual on the International Law Applicable to Cyber Warfare" (Schmitt, 2017, p. 1). More recently, additional rules have been added that now give us the Tallinn Manual 2.0. Tallinn 2.0 acknowledges the difficulty of applying *jus ad bellum* to cyber operations. In "the cyber context, it is not the instrument used that determines whether the use of force threshold has been crossed, but rather . . . the consequences of the operation and its surrounding circumstances" (Schmitt, 2017, p. 328). For example, what if a "just" cyber-attack is initiated by Country A into Country B, but the cyber-attack does not stop there. What happens when this attack bleeds over into Country C, Country D, etc.? Can those third and fourth countries legally respond and retaliate against Country A? Country A may have set up the attack solely against Country B and done everything in its power to limit the scope to Country B alone. However, cyber-attacks can clearly go beyond the intended target, as evidenced by major cyber operations like StuxNet, which spread extensively beyond its apparent target of Iran's nuclear facilities, despite a number of control measures programmed into the weapon (Raymond et al., 2013). The scope and scale of current cyber-attacks around the world show us the importance of just war theory to international cyber issues.

The second component of just war theory is *jus in bello*. As J. Moussa elaborates, "*Jus in bello* . . . has as its aim the conciliation of 'the necessities of war with the laws of humanity' by setting clear limits on the conduct of military operations" (Moussa, 2008, p. 965). The criteria of *jus in bello* include distinction (also called discrimination), proportionality, military necessity, fair treatment of any prisoners of war, and just means (Pattison, 2009, p. 367). The first element of distinction or discrimination involves the idea that the actor using force cannot indiscriminately use such force (Pattison, 2009, p. 367). Legitimate (proper military combatants) and illegitimate (civilian non-combatant) targets must be distinguished. This is a difficult metric in cyber operations as a cyber weapon may (intentionally or unintentionally) affect the computer systems of civilians or other noncombatants. Another criterion involved with *jus in bello* is proportionality. As J. Pattison writes, "The use of force must be proportionate to the military advantage gained. The excessive use of force against combatants is prohibited" (Pattison, 2009, p. 367). But what about use of force in a cyber operation? If Country A's cyber operations are attacked by Country B, is Country A only allowed to respond in a similar manner involving cyber-attacks? If a country is only allowed to respond in like kind via cyber, what happens if an aggressor is much less technologically dependent than the responding country? The answers to these types of questions will remain unclear until the international community comes up with a generally-accepted consensus on such issues.

Three other criteria for *jus in bello* involve military necessity, the fair treatment of prisoners of war, and just means (Pattison, 2009, p. 367). An important point under these criteria is that there is a "prohibition on the use of certain weapons and methods, such as biological warfare and anti-personnel mines" that are considered *malum in se*, or evil in itself (Pattison, 2009, p. 367). Therefore, the saying "all is fair in love and war" is not true; there are limitations on the types of warfare that can legitimately be used and that will be accepted internationally. But are there any limitations on cyber-attacks? At what point will a cyber-attack be deemed as "going too far" by the international community? As more and more cyber-attacks happen, and as more become publicized (many cyber-attacks will never be known by the general public), perhaps greater clarity around these questions will evolve.

Colonel James Cook, when discussing just war theory, argues that the theory "passes judgment of the *effects* rather than the means or media of aggression" (Cook, 2010, p. 412). Just war theory is most useful when those effects meet three conditions: "(a) [they] must be known or strongly predicted; (b) they must be linked to known intentional actors; and finally, (c) they must manifest as destruction or imminent destruction—of lives, property, governments, cultures, etc." (Cook, 2010, p. 412). Just war theory often cannot be easily applied to cyber weaponization because the effects are uncertain or the actors cannot reliably be identified. Evidence may strongly link one nation to an attack when, in fact, it was another actor altogether. However, meeting these three conditions makes it much easier to apply just war theory to cyber operations.

Cyber-Kinetic Attacks

Cook's third criterion, destruction of life, property, and the like, was for decades considered anathema to cyber operations, due to the use of cyber primarily in espionage/intelligence and information operations. StuxNet's appearance in 2007, however, demonstrated once and for all that cyber weapons could be used for kinetic effect—destroying many of Iran's nuclear centrifuges, as well as unintentionally damaging systems in Europe and elsewhere (Raymond et al., 2013). In more than a decade since StuxNet, the Internet of Things (IoT), including the expanded use of "smart" grid Internet-connected infrastructure and smart meters for electric, water, gas and other utilities, self-driving cars, and even network-controlled medical devices, has made cyber-kinetic attacks both more achievable and potentially much more devastating (Payne & Abegaz, 2018).

Cyber has already been widely adopted as a component of hybrid warfare, as seen in Iran's response to the Green Movement in 2009, or Russia's engagement with separatists in eastern Ukraine in 2014 (Duggan, 2015). The potential for cyber-kinetic warfare has been demonstrated, though, in which cyber weapons alone target critical infrastructure (Parks & Duggan, 2011; Raymond et al., 2013). In a proof of concept, a US national laboratory demonstrated a successful attack against a nuclear power plant in the US as far back as 2007 (Meserve, 2007), the same year that StuxNet damaged the Iranian nuclear centrifuges (Raymond et al., 2013). It should be noted, though, that while cyber weapons like these can be deployed without the use of traditional kinetic weapons (bullets, bombs, etc.), human intelligence operatives may still be a key in getting past layered defenses, such as air-gapped systems that are not connected to the public-facing Internet—a common example is the scattering of infected USB thumb drives in or near classified facilities (Beaumont, 2010).

Under Cook's standard of judging the effects rather than the means, a cyber-attack that shuts down a power grid, sabotages a traffic system (whether ground or air), tampers with water treatment or other critical infrastructure, could arguably be deemed an act of war if the end result were as damaging as a kinetic projectile or bomb on the same infrastructure. A computer virus or ransomware that shuts down medical devices, like pacemakers or insulin pumps (or that overloads pacemakers and morphine pumps), could be held to the same standard as if the aggressor had fired bullets through the affected victims.

However, the issue of determining who fired a weapon in the physical world is usually resolved using surveillance videos, interviewing eye witnesses near the scene, examining fingerprints and explosive residue, and the like. Trying to identify the culprit in a cyber-attack circles us back to the problem of attribution. No videos, no physical trace evidence, no eye witnesses may exist near the site of the attack. And, gathering and reviewing the scarce electronic evidence, from poring over network and system logs to reverse-engineering any malware involved, may take weeks, months, years, or never be possible to determine with certainty at all.

A physical missile launch leaves satellite videos, a vapor trail, and has a trajectory, while a cyber weapon may sit undetected for months before a kinetic effect takes place and be difficult or impossible to investigate for years after the attack is over. An additional complicating issue may be the fact that roughly 85 percent of US critical infrastructure is privately owned, by communications companies, power corporations and cooperatives, banks, and so on (FEMA, 2011). Does an attack on privately-owned corporate equipment through privately-owned networks by a nation-state or alleged terrorist organization merit consideration as a matter of national security? Possibly, if the effect (economic sabotage, loss of life, etc.) is significant enough, but once again, the attribution problem may never be completely resolved.

Toward Deterrence in Cyber

So, then, how do we deter a threat that we can't see, aimed at or through privately-owned assets, from an enemy we can't readily determine, using weapons we might not be able to control?

A partial answer might be taking action not exclusively through cyber. As with any attack, diplomatic, informational, military, economic, and legal channels may be employed to punish a rogue actor. In cases of cybercrime, particularly, legal action has been successful in a number of cases, including suing for damages, filing criminal charges against individuals and organizations. Informational approaches, including "naming and shaming" cyber actors and government sponsors as seen in the FBI's Most Wanted posters featuring Russian and Chinese hackers and government units, have proven successful in leading to the arrest and indictment of a number of key malicious hackers. Diplomacy may have worked in the case of former US President Obama's cyber agreement with Chinese President Xi in 2015, after which alleged economic cyber-espionage appeared to drop dramatically. Economic sanctions and even military action may become necessary in some cases, but threats remain.

Part of the answer to cyber threats would seem to require cyber-specific deterrence through and in the cyber domain itself. Further, deterrence is not based solely on the likelihood of reprisal, as noted by Snyder, but also on the threat of denial (Snyder, 1961). A better question perhaps, then, becomes how we can best deny or stop such attacks before they ever take shape.

In addition to military, legal, economic, and diplomatic forms of deterrence by punishment, we propose greater deterrence by denial. In particular, we propose the creation and maintenance of a well-trained, ready force of cyber heroes capable of both defense and offense, deployed to secure public and private computer systems and networks, in roles across military, government, and private industry.

Proposal: Civilian Cyber Corps

One of the obstacles in protecting our national infrastructure has become the issue of sheer workforce numbers. As many as

290,000 job openings in cybersecurity go unfilled each year in the United States alone (Cyber Seek, 2018), with a predicted worldwide shortage totaling in the millions over the next few years.

In order to maintain layered defenses, actively detect intrusions, and develop state-of-the-art offensive capabilities, the US and its allies require an adequately trained workforce. The authors propose, as one way of addressing the critical shortage of cyber guardians, the establishment of a Civilian Cyber Corps. Similar to US National Guard and Reserve components, participants could spend one weekend a month and two weeks a year training in cyber and protecting government and critical cyber infrastructure, while working as civilian employees in government or private industry. Employers would support and make allowances for members of the corps to participate in training, and hold positions when members are deployed, just as with reservists. This type of civil-military cooperation has proven crucial both in times of peace and during periods of armed conflict.

Training qualified civilians in the latest defensive capabilities, and, with sufficient clearance, offensive tactics would, at the same time, respond to the critical need for cybersecurity professionals in private industry and government, and sustain a corps of well-trained cyber heroes to protect and restore critical systems nationally. As is the case with National Guard and Reserve units within each military service, such a force would also have the capacity to support defensive and offensive operations against adversaries should the need arise to "deploy" or "activate" the civilian corps during times of war or national crisis. This approach aligns with the more modern notion of "citizens-who-become soldiers" proposed by Dubik (2016), as the government protects and defends its civilians and private institutions through cyber training, and the civilians in turn protect national security interests via that training.

One key to garnering broad participation would be the civilian nature of the corps. Such elements as lower fitness requirements and higher age allowances would be recommended for cyber recruits. High-tech industry cultural components, like T-shirts and tennis shoes instead of boots and fatigues, could beckon multiple generations of capable persons across race, gender, socio-economic status, and even persons with physical disabilities who wish to serve their country into the Civilian Cyber Corps. A corps of this nature could be a bridge to high-paying jobs, especially for the economically disadvantaged and for returning/retiring armed service personnel.

The establishment of a Civilian Cyber Corps may be more immediately important for national security in another sense, as well. A 2017 McKinsey report predicted that as many as 73 million jobs may be partially or fully automated by 2030 due to robots, apps, drones, AI assistants, and related automation technologies. The jobs noted in the report included both low-skill and "middle-skill" jobs, including data processing and other typically white-collar career fields, noting that "these activities make up 51 percent of activities in the economy, accounting for almost $2.7 trillion in wages."

In the most dystopian case, a majority of jobs could be partially or fully automated, resulting in an entirely new kind of Great Depression in the mid-twenty-first century. During the last Great Depression in the 1930's, more than 20 percent of the US population became unemployed. At that time, New Deal legislation like the creation of the Civilian Conservation Corps (CCC) helped put hundreds of thousands of people to work on projects that built infrastructure, dams, roads, power generation, and more. The new CCC proposed in this work could build and protect new kinds of cyber, physical, and technological infrastructure and stand ready to defend the nation for the next generation and beyond.

Conclusion and Future Work

The establishment of a Civilian Cyber Corps would require significant funding, perhaps at the level of the last century's New

Deal, but starting before widespread unemployment or other crisis occurs could temper the magnitude of such spending. And, while the cost of such an undertaking will be high, this approach addresses both cyberwarfare and criminal cyber acts by preparing better-trained cyber professionals in both government and industry and could be offset by savings and cost avoidance realized by preventing and reducing the impact of cybercrime.

Similar to Israel's "Iron Dome" defensive systems, stronger government cyber defenses would protect private citizens and assets through better-defended networks and systems at the perimeter and throughout our national infrastructure. Further, the significant increase in the number of cyber defenders and cyber operators would serve as an additional deterrent by denial, due to the sheer quantity of cyber guardians actively protecting and patrolling our critical systems and networks.

The corps could be a valuable career option for unemployed/under-employed citizens, including those from displaced industries due to automation by artificial intelligence and related technologies, as well as a career starter and patriotic service opportunity for a new generation of cyber heroes.

In addition to allocating funding and gathering the national will to support such an effort, other limitations exist that must be explored further. For example, this approach does not directly address cyber espionage. While more adequately- and actively-trained cyber professionals in government and industry would be better prepared to thwart and detect theft of sensitive data and intellectual property, subtle cybercrimes such as these would continue to be a threat to economic prosperity and stability. Unfortunately, there is no generally-accepted "just espionage theory" as with war theory, and, quite the opposite, nations are generally expected to gather intelligence to protect their interests.

Military, legal, economic, and diplomatic approaches to deterrence against cyber-attacks remain highly valuable, but stronger, smarter, better-trained cyber professionals in government and private industry mean stronger defenses for a more secure nation, and more effective offense in the event of hybrid conflict or actual war.

References

Beaumont, P. (2010). Stuxnet worm heralds new era of global cyberwar. *The Guardian*. Retrieved October 2, 2017 from https://www.theguardian.com/technology/2010/sep/30/stuxnet-worm-new-era-global-cyberwar

Clarke, R. A., & Knake, R. K. (2014). *Cyber war*. Tantor Media, Incorporated.

Cook, C. (2010). 'Cyberation' and Just War Doctrine: A Response to Randall Dipert. *Journal of Military Ethics*, 411-423.

Crisher, B. (2005). Altering Jus Ad Bellum: Just War Theory in the 21st Century and the 2002 National Security Strategy of the United States. *Critique: A worldwide journal of politics*, 1-30.

CyberSeek.org. (2018). Cybersecurity Supply/Demand Heat Map. Retrieved January 20, 2018 from http://cyberseek.org/heatmap.html

DiMeglio, M. P. (2005). The Evolution of the Just War Tradition: Defining Jus Post Bellum. *Military Law Review*.

Dubik, J. M. (2016). *Just War Reconsidered: Strategy, Ethics and Theory.* Lexington, KY. University Press of Kentucky.

Duggan, P. M. (2015). Strategic Development of Special Warfare in Cyberspace. *Joint Forces Quarterly 79*(4), 46-53.

FEMA. (2011). Critical Infrastructure: Long-term trends and drivers and their implications for emergency management. Strategic Foresight Initiative, FEMA. Retrieved October 3, 2017 from https://www.fema.gov/pdf/about/programs/oppa/critical_infrastructure_paper.pdf

Geers, K. (2010). The challenge of cyber-attack deterrence. *Computer Law & Security Review 26*(3), 290-297.

Goodman, W. (2010). Cyber Deterrence. Tougher in Theory than in Practice? *Strategic Studies Quarterly*, Fall 2010.

Lango, J. W. (2005). Preventative Wars, Just War Principles, And The United Nations. *Journal of Ethics*, 247-268.

May, L., & Delston, J.B. (Eds.). (2016). *Applied ethics: A multicultural approach*. New York, NY. Routledge.

Moussa, J. (2008). Can jus ad bellum override jus in bello? Reaffirming the separation of the two bodies of law. *International Review of the Red Cross*.

National Catholic Reporter (2017). *Experts discuss just war theory and cyber-attacks.* Retrieved August 2, 2017 from National Catholic Reporter web site at: https://www.ncronline.org/news/justice/experts-discuss-just-war-theory-and-cyber-attacks

Parks, R. C. & Duggan, D. P. (2011). Principles of Cyberwarfare. *IEEE Security & Privacy 9*(5), 30-35. DOI: 10.1109/MSP.2011.138.

Pattison, J. (2009). Humanitarian Intervention, the Responsibility to Protect and jus in bello. *Global Responsibility to Protect 1*.

Payne, B. R., & Abegaz, T. T. (2017). Securing the Internet of Things: Best Practices for Deploying IoT Devices. In El-Sheikh, Ertual, Francia, and Hernandez (Eds.) *Computer and Network Security Essentials*, pp. 493-506. Springer, Cham, Switzerland.

Raymond, D., Conti, G., Cross, T. & Fanelli, R. (2013). A control measure framework to limit collateral damage and propagation of cyber weapons. In Podins, K., Stinissen, J., & Maybaum, M. (Eds.), *2013 5th International Conference on Cyber Conflict* (CyCon). NATO CCD COE, Tallinn.

Schmitt, M. N. (2017). *Tallinn Manual 2.0 on the International Law Applicable to Cyber Operations.* Cambridge: Cambridge University Press.

Snyder, G. H. (1961). *Deterrence and Defense: Toward a theory of national security.* Princeton Legacy Library.

Walzer, M. (1977). *Just and Unjust Wars*. New York: Basic Books.

Williams, Jr., R. E., & Caldwell, D. (2006). Jus Post Bellum: Just War Theory and the Principles of Just Peace. *International Studies Perspectives*, 309-320.

2

South Africa and the Cyber Warfare Threat: A Strategic Overview

Noëlle van der Waag-Cowling

"Both state and non-state actors understand that the cyber domain favours offensive action. Static defences are the modern equivalent of a Maginot Line: vulnerable to incessant battering by an unknown opponent and easily circumvented by manoeuvre."[1]

Abstract

South Africa experiences challenges in terms of cyber threats within the African digital space. Foremost amongst these is the framing of a comprehensive national cyber strategy and further to that, a cyber warfare strategy. There are key questions which arise around such strategies in terms of governance, policy development, doctrine, capability development, knowledge collaboration, diplomatic posture and the sharing of information. Furthermore, there is a need to make rapid progress within the human capital environment. While defence planners do not foresee a conventional short- or medium-term military threat, the same cannot be said for cyber and possibly cyber/kinetic attacks due to the global nature of cyber threats and the proliferation of possible adversaries. From a South African perspective this presents a complex backdrop against which cyber strategies must be framed.

---

1 Paul Cornish et al, "On Cyber Warfare." *A Chatham House Report*, (November 2010) : 21.

Introduction

The ever growing threat of cyber-attacks and cyber warfare on nation states[2] means that globally, states are diverting considerable resources to combating cyber-related offensives.[3] South Africa too has identified the need to mitigate the cyber threat; however, despite certain legislative prescripts having been passed, the country lags behind in terms of strategic planning, the implementation of effective cyber entities and an effective programme to build cyber resilience in the population.[4] [5]

According to the International Telecommunications Union's (ITU) 2017 Global Cyber Security Index Report, South Africa ranks fifty-seven on the Global Cyber Security Commitment Score and 6th in Africa behind Mauritius, Rwanda, Kenya, Nigeria, and Uganda.[6] In terms of national legislation, cyber matters are currently informed by the National Cyber Policy Framework (NCPF), which provides an outline of the role of certain government departments in combatting cyber threats. However, it is not an in-depth document nor a strategy and this creates difficulties in terms of defining roles, responsibilities, and a clear way forward.

Africa as a whole, experiences challenges in terms of threats within its digital space.[7] While South Africa is one of the leaders

2 See for example: Uche Mbanaso, "Cyber warfare: African research must address emerging reality." *The African Journal of Information and Communication (AJIC)*, (18) (2016): 157-164; Johan Sigholm, J. "Non-State Actors in Cyberspace Operations." *Journal of Military Studies*, 4(1) (2016): 1-37. And Jones, A. and Kovacich, G.L. "Global Information Warfare: The New Digital Battlefield." (Second Edition, Taylor and Francis, USA), (2016).

3 See for example: George Manson, "Cyberwar: The United States and China Prepare for the Next Generation of Conflict." *Journal Comparative Strategy*, Volume 30, 2, (2011): 121-133 and Tatar, Ünal et al. "*A Comparative Analysis of the National Cyber Security Strategies of Leading Nations*." International Conference on Cyber Warfare and Security; Reading: 211-X. Reading: Academic Conferences International Limited. (2014)

4 "National Cyber Policy Framework for South Africa." State Security Agency, South African Government Gazette No. 39475 of 4 December (2015); "DOD Planning Instruments for 2015 to 2020." South African Department of Defence, (2015) and "South African Defence Review 2015." South African Department of Defence, (2015) and "Cybercrimes and Cybersecurity Bill." (proposed section 75), (2017) Republic of South Africa.

5 Brett van Niekerk, "An Analysis of Cyber-Incidents in South Africa." *The African Journal of Information and Communication* (AIJC), (20) (2017): 115.

6 United Nations. "Global Cybersecurity Index (GCI)" International Telecommunications Union (2017): 15.

7 UN, "Global Cybersecurity Index," 15.

in this arena, a number of substantial challenges remain; foremost amongst these is the need for a national cyber strategy, and further to that, a cyber warfare strategy. There are a number of key questions in the South African context around such strategies in terms of governance, policy development, doctrine, capability development, knowledge collaboration, and the sharing of information with other states. Associated questions regarding which government entity is responsible for securing South Africa's digital landscape are equally difficult to answer.

The requirement for rapid progress in both the human capital and technological environment is self-evident. South Africa is home to the most advanced economy in Africa, a service hub for the entire region and an important regional power. While defence planners do not foresee a conventional medium-term military threat,[8] the same cannot necessarily be said of cyber and possibly cyber/kinetic attacks due to the global nature of such threats and the proliferation of non-state actor adversaries. Aside from the need to secure the South African National Defence Force (SANDF) from cyber-attacks and espionage, the protection of both national critical infrastructure as well as the South African arms manufacturing industry, both from a military as well as an economic perspective, are equally important considerations.

There are also a number of foreign policy challenges for South Africa in terms of cyber cooperation as the country is a member of a number of international organizations from the United Nations (UN), the African Union (AU), the Southern African Development Community (SADC), Brazil –Russia – India – China – South Africa (BRICS), and the Commonwealth of Nations.[9] This is set against a backdrop of the African continent which, as a whole, faces many

---

8 "South African Defence Review" (2015): 3-14.

9 See for Example: Abel Esterhuyse, "The South African Threat Agenda: Between Political Agendas, Perceptions and Contradictions." *S&F Sicherheit und Frieden*, (2016) (Seite): 191 – 197; and Theo Neethling, "South Africa's Foreign Policy and the BRICS Formation: Reflections on the Quest for the 'Right' Economic-diplomatic Strategy," *Insight on Africa*, (2017), Vol 9: 39 – 61.

security risks combined with outdated resource shortfalls in the Information Communication and Technology (ICT) sector and significant demands in terms of building the required human capital in the cyber domain. Further to this, Africa's number of internet users is exploding as is its physical cable connectivity to other continents.[10] [11] It is therefore perhaps fair to assume that the South African threat landscape is large and somewhat exposed. From a national perspective, this presents a complex backdrop against which cyber strategies must be framed. The need for collaboration in terms of knowledge building and intelligence sharing in the cyber domain cannot be overstated and policy formulation in this regard will present a challenge to defence planners.

This paper explores the multi-faceted political and organizational factors which will influence the design and nature of a cyber warfare strategy for South Africa and proposes some key considerations, requirements, and outcomes to that end. Issues of a technical or technological nature are deliberately not discussed at this juncture for the sake of brevity as the intention is to place the focus of the discussion on a strategic level.

Cyber Diplomacy and Cyber Power—the South African Context

From a military and strategic viewpoint, the key questions are where does South Africa position itself in terms of both strategic alliances and security threats and, more immediately, how will this influence its cyber warfare strategy?

From a strategic cooperation perspective South Africa is firmly positioned as a leading power on the African continent and the national stance is often echoed in the statement "African solutions to African problems."[12] As such, it firmly positions itself within the

10 Aurthur Goldstuck, "Massive SA growth of 4G, Wi-Fi, in next five years." *Mail and Guardian*, 22 Apr 2015. https://mg.co.za/article/2015-04-22-massive-sa-growth-of-4g-wi-fi-in-next-five-years.

11 Van Niekerk, "An Analysis of Cyber-Incidents in South Africa." 115.

12 Hussein Solomon, "African Solutions to Africa's Problems? African Approaches to Peace, Security

ambit of the AU, while, closer to home, South Africa is the leading state in SADC, where defence cooperation constitutes one of the twenty-seven legally-binding SADC protocols.

South African foreign policy is largely predicated upon its regional objectives in Africa. The country's posture is frequently criticized for appearing to be aimless at best and duplicitous at worst and sets the tone for what some perceive as a largely indifferent relationship with the West. In spite of this perception, some scholars provide assisted insight; Beresford provides a key to dissecting the foreign policy stance:

> If they are able to see beyond the fiery rhetoric, Western policy makers are likely to find in South Africa a potentially pliant partner whose core foreign policy interests in Africa are not all that dissimilar from the Western powers: stability and economic opportunity. To do so requires a detailed understanding of South Africa's transition to democracy, and how the constellation of ideas, identities and ideologies that emerged from it to forge what can be identified as South Africa's approach to tackling Africa's security dilemmas. [13]

In general, African states are wary of the presence of large western powers on the Continent. The same does not appear to apply to certain countries from the East, with China being a prime example. This caution with regards to the West stems in part to past colonial relationships but extends further to the military presence of these powers in Africa and the ever-increasing specter of proxy forces operating within the continent's many conflict zones. Asian countries, on the other hand, are viewed as robust trading partners with little interest in sovereign political issues and ones who are investing substantially in infrastructure on the Continent. This

---

and Stability." *Scientia Militaria - South African Journal of Military Studies*, V 43, N 1, (May. 2015): p 58.

13 Beresford, "A responsibility to protect Africa from the West? South Africa and the NATO intervention in Libya." *International Politics*, Vol 52 (93), (2015): 290.

presents something of a dilemma for South Africa, whose largest export trading partners are Germany and the United States (US) with China in third place.

It is becoming increasingly clear that South Africa is aligning itself firmly within BRICS and eschewing its historical ties with the West. This speaks to the South-South multilateralism which forms an important cornerstone of her foreign policy.[14] While initially China presented itself as South Africa's major trading partner within BRICS, recent developments indicate that the relationship between Russia and South Africa is becoming one of primary importance both strategically and economically. From a cyber standpoint, it is notable that the two have now entered into a formal cyber security agreement, which was signed in September 2017. According to the Russian foreign ministry, the "pact demonstrated Russia's and South Africa's commitment to expanding bilateral cooperation in one of the most up-to-date areas of international and national security."[15] Importantly, the Russian statement stressed the strategic nature of this agreement. The pact will ostensibly be given instrumentality through the creation of a joint threat response system, a research programme, and specialist training initiatives.[16] Looking to the broader BRICS community, it is suggested that South Africa would do well to pursue similar types of cyber-related agreements with India, which is something of a technological hegemon.

South Africa's relations with the United States are more difficult to dissect. While trade, cultural, and academic relations between the two countries remain strong, the overall diplomatic bilateral relationship, while cordial, appears less warm than during the Mandela era. There are a number of historical and current factors which contribute to this, most notably ideological differences.

---

14 Theo Neethling, "South Africa and AFRICOM: Reflections on a lukewarm relationship." *South African Journal of International Affairs*. Vol 22: 1, (2015): 115.

15 Tass Russian News Agency. "Russian, South African top diplomats ink cyber security cooperation deal." 4 September 2017, http://tass.com/politics/963564.

16 Ibid.

The establishment of United States Africa Command (AFRICOM) in 2007 appears to have contributed to the complexity of this relationship. Greg Mills assesses the situation rather strongly and writes:

> South Africa's relationship with international actors is another potential force multiplier in Africa. However, this has been hamstrung, to an extent, by the South African government's schizophrenic relationship with the United States: perceived as a major trade and investment partner on the one hand, and with paranoia about imperial intentions on the other, viz. the hullabaloo over the creation of AFRICOM.[17]

AFRICOM appears to have struggled to gain traction in Africa and most particularly in Southern Africa where South Africa has been a leading voice in terms of SADC rejecting any US attempts to establish a locality within the region. Key to this is a fear of a hyper power establishing a substantial military force within domestic striking range; however, there are other less-obvious concerns, for example, the establishment of proxy forces within the region, which appear to be multiplying. Despite this, military cooperation has continued and just recently the two nations participated in a major conventional exercise named Shared Accord at the South African Army Combat Training Centre.[18]

Despite the ongoing yet cautious military relationship between the US and South Africa, there is nevertheless a seemingly-conspicuous absence of any cyber-related collaboration, and it is difficult to gauge to what extent the two countries cooperate in terms of sharing intelligence. Both of these factors could prove to be a serious setback to South Africa in terms of cyber skills development, combating terrorist threats, and organized crime.

---

17 Neethling, "South Africa and AFRICOM: Reflections on a lukewarm relationship." 112.

18 Stand-To. The Official Focus of the US Army. "Shared Accord - (2017). https://www.army.mil/standto/2017-07-17.

It is argued that South Africa cannot afford to ignore potential opportunities to expand its knowledge and capability base, which could result from a more extensive partnership with the US in the cyber domain. The US is a major military and intelligence cyber power but, in addition, it also remains the global hub of technology innovation. US ICT technologies are utilized extensively in South Africa and this factor alone should underline the need for deeper interchange.

Likewise, South Africa's relations within the Commonwealth appear to lack much political or diplomatic substance. This is of particular import if one considers that South Africa is the only one of the four former British Dominions which has never been included in the Five Eyes Agreement, despite its initial inception being as early as 1948.

Considering the bilateral British-South Africa partnership more specifically, this is somewhat hindered by the United Kingdom's (UK) former position as South Africa's colonial hegemon. A limited level of military interaction continues through training and education exchanges and certain limited bilateral exercises.[19] Nonetheless, the UK's African military focus has perceptibly shifted towards East Africa and the "Arc of Instability." This should nevertheless not detract from the necessity for cyber and intelligence cooperation. The existing bonds between the two entities is evidenced by the vast number of people who travel between South Africa and the UK, as well as huge financial flows due to trade and South African companies with dual listings on the London Stock Exchange.[20] Given Britain's acknowledged expertise in policing, intelligence, and combating terrorism, as well as the considerable resources it is committing to its national cyber efforts at Government Communications Headquarters (GCHQ), the somewhat distant relationship between the two former

19 Joint Communiqué on the occasion of the 11th Meeting of the United Kingdom – South Africa Bilateral Ministerial Forum, London, 19 October 2015. http://southafricahouseuk.com/documents/11bilatrlfrm.pdf.

20 Alex Vines et al "UK and South Africa: A Relationship Worth Maintaining." Chatham House, (2013), https://www.chathamhouse.org/media/comment/view/194212.

allies is problematic. More so if one considers that cyber cooperation would be mutually beneficial given the documented movements of certain terrorist operatives between the two countries.

Within the overall cyber diplomacy realm, it is important to note that cyber also presents strategic opportunities. Through the development of advanced cyber capabilities, small and medium size powers have the prospect of increasing their military power beyond the physical size of their armed forces. Nye points out that "The digital domain is characterized by power diffusion" and that "the characteristics of cyber space reduce some of the power differentials amongst the various actors."[21] This implies increased potential for cyber (military or state) power projection capabilities for South Africa.

Key Focus Areas of South Africa's Cyber Warfare Strategy

Moving on to the strategic landscape, there are a number of considerations. The NCPF[22] appears to be a rather granulated approach at best or an overly-fragmented approach at worst. Various government departments have been accorded certain roles; however, these appear to overlap and are highly dependent on the sharing of information and being properly resourced, which could prove to be problematic. On a military level, defence planning indicates that establishing cyber capabilities is one of three top defence priorities over the medium term.[23] [24] Key to this is the formulation of the cyber warfare strategy, which is due to be presented to Parliament. Of

21 Joseph Nye, "Cyber Power." Monograph, Harvard University Belfer Center for Science and International Affairs. May, (2010): 1.

22 "National Cyber Policy Framework for South Africa." (2015).

23 "DoD Annual Performance Plan" (2016): 26.

24 "Level of Implementation of the Cyber Warfare Plan. This will be conducted through a phase approach, which is as follows: Phase 1 – Establish HQ; Phase 2 – Finalise functions; Phase 3 – Finalise structures; Phase 4 – Obtain budget; Phase 5 – Establish capabilities; Phase 6 – Create cyber awareness program. Establishment of a Cyber Command Centre, which must be fully operational by FY2018/19." DoD Annual Performance Plan, (2015), 121.

some concern is the fact that Defence is currently lagging behind the set target dates to formulate and present such a strategy.[25] This may partly be due to the rather nebulous nature of the NCPF but also points to the lack of literature, data, and knowledge on cyber warfare within the South African context.

So what are the key focus areas of South Africa's cyber warfare strategy? As far as the National Security Strategy (NSS) is concerned, South Africa states that its national security:

> includes dimensions of domestic, regional and continental national security. National interests will continue to drive the involvement of major powers in Africa, especially where vital interests are at stake. While 'traditional' interests such as oil and strategic minerals remain important, the perceived threat posed by Islamist extremism to intra-state security is becoming increasingly more important. The emerging norms for the coalition of the willing may in future require the Republic of South Africa (RSA) to become militarily involved. Globalisation creates opportunities for transnationally operating aggressive non-state actors that make use of global cyber networks, financial networks and transportation networks, especially when terrorist groups and criminal groups have a common cause. The nodal points—seaports, airports, computer servers, banks–in all these networks form the centres of gravity of their operations, with the ultimate end-state to bring about political instability.[26]

The NSS points to the importance which strategic planners in the security cluster place on cyber threats. While South Africa ranks as a medium-strength global military power, the armed forces currently faces numerous challenges, not least of all a drastically-reduced budget, which now stands at under 1% of Gross Domestic

25 "DOD Planning Instruments for 2015 to 2020," 3.

26 DOD Annual Performance Plan, 2-3.

Product (GDP). Notwithstanding this, the development of a cyber-warfare capability is a priority within the Medium Term Strategic Framework (MSTF).[27] The 2015-2020 Defence Planning Instrument states that

> The Department of Defence (DoD) will develop a comprehensive Cyber Warfare Strategy aligned with the national policy regarding South Africa's posture and capabilities related to offensive information warfare actions.[28] Furthermore, Defence has been tasked to contribute towards capacitating a Cyber-Security Institution by establishing a Cyber Command Centre Headquarters.[29]

There are some key considerations which need to be incorporated into the design of such a strategy. It is imperative for the architects of the strategy to remain mindful of the three key elements of strategy in this process: "A national security strategy, like any strategy, must be a combination of ends (what we are seeking to achieve), ways (the ways by which we seek to achieve those ends), and means (the resources we can devote to achieving the ends)."[30]

Firstly, it is postulated that when initiating a cyber warfare strategy, the overall focus should be on the national cyber security architecture and not narrowly focused on military requirements or the broader security cluster. South Africa has an overall moderate cyber dependence. This varies somewhat between sectors where, for example, the private financial, energy, and medical sectors may have a high dependence, while agriculture may have a lower dependence. The governmental sectors probably rest on a moderate-dependence threshold; even so, any widespread and sustained cyber attack would likely cause potential chaos. The key

27 DOD Planning Instrument 2015-20, 55.

28 DOD Planning Instrument, 9.

29 DOD Planning Instrument, 13.

30 Cornish. "On Cyber Warfare." 25.

issues for state planners must be in ensuring the ability of the State and the to economy remain functional. As Robinson et al point out:

> In the aftermath of cyber warfare, states may find difficulty in bringing cyber dependent infrastructure back online, and this has the potential to weaken state authority and subsequently delay the return of peace and security in the region. In extreme cases, a government's inability to resume the provision of basic services could lead to state collapse.[31]

One only has to consider the growing violence of regular service delivery protests in the country to envision the potential reality of such a scenario.[32] Mitigation and recovery strategies in the event of such an attack therefore require thoughtful and detailed planning and testing. For the same reasons, South Africa needs to assist its neighbours in cyber defence, given the knock-on effects of possible instability.

Secondly, from a notional point of view, it is postulated that South Africa's cyber warfare strategy should be defensively minded. The principal objective should be to protect the populace, the economy, and the DoD itself. In this sense, the strategic emphasis would fall upon practicing due diligence and employing the latest international standards in cyber defence.[33] The development of advanced forms of offensive cyber capability would be a second priority and possibly developed in conjunction with identified allied forces within the cyber domain. The notion of working with other states is cardinal both in terms of pooling expertise and leveraging access to greater resources as well as in strengthening

---

31 Michael Robinson et al, "An Introduction to Cyber Peacekeeping." https://arxiv.org/pdf/1710.09616.pdf, (Oct 2017): 18.

32 Mathew Savides, M. "Fewer protests in 2016, but they were more violent." https://www.timeslive.co.za/news/south-africa/2017-02-01-fewer-protests-in-2016-but-they-were-more-violent/.

33 Emilio Iasiello, E. "Is Cyber Deterrence an Illusory Course of Action?" *Journal of Strategic Security*, 7, No. 1 (2013): 67.

the principle of collective security. The principle role of such a capability should be to provide credible deterrence.

The question of cyber deterrence is admittedly fraught with many issues—most notably attribution and issues pertaining to Conventions on the Law of Armed Conflict (LOAC). The first task is to ensure that deterrence is in fact credible, which implies developing a means of retaliation that is sufficiently severe to discourage attack in the first place. Convincing both state and non-state actors of the veracity of this ability is dependent on having established a track record and body of work within a nation state, which gives credence to the belief that such capability and agility exists and is functional.[34] Regarding legal issues, it is fairly obvious that any cyber warfare strategy must pay sufficient attention to the downstream development of policy and doctrine pertaining to the issue of *jus bellum iustum*. The question of a Just War and what constitutes an act of war[35] with regards to cyber attack remains the subject of ongoing legal debate and theoretical development. Suffice to say that the usual legal conventions and provisions which surround state conflict cannot simply be applied *mutatis mutandis* to cyber acts of aggression, primarily because the aggressor's identity is usually concealed. Iasiello underscores the critical importance of reacting appropriately,

> Here, a nation state's credibility is interlinked with proportionality in that the nation state must not only strike back against the aggressor but it must do so in a way as to make its point—that is, it must be a forceful strike—but not so forceful as to solicit negative reaction in the global community.[36]

Proportionality essentially entails a weighted response which is both commensurate with the initial attack and yet avoids escalation.

34 Iasiello. "Is Cyber Deterrence an Illusory Course of Action?" 57.

35 Cornish, "On Cyber Warfare." 13.

36 Iasiello, "Is Cyber Deterrence an Illusory Course of Action?" 60.

Having said that, blowback is always a consideration, whether by means of a counter-retaliation or diplomatic censure.[37]

This raises the matter of strategic emphasis. As is the case with most states, South Africa is highly dependent on national critical infrastructure (NCI) and communications infrastructure. This is one of the few countries in Africa with highly-developed infrastructure, particularly within the ambit of the energy, communications, and transport sectors. This is a murky area for defence planners, as the responsibility for NCI appears to fall somewhere between the State Security Agency, the SA Police Service, and Defence.[38] Suffice to say that kinetic attacks on infrastructure form part of warfare and, while not illegal in terms of the law of war, they do constitute an act of war. The potential for cyber kinetic attacks indicates that responsibility for NCIs should ultimately rest with Defence and it is rather crucial that the authorities revisit this issue and delineate clear lines of responsibility.[39]

Finally, human capacity building with regards to both training and recruitment require significant attention. The global shortage of cyber skills is well documented. Arguably, this presents even more of a challenge in countries such as South Africa, which tend to experience high levels of skilled emigration. Defence will therefore face some challenges in building a skilled and elite cyber warrior force and, indeed, in retaining such personnel after having invested heavily in them. One of the areas which will require due attention is the development of a capable technical workforce. South Africa will need to avoid creating a reliance on existing tools

37 Iasiello, "Is Cyber Deterrence an Illusory Course of Action?" 60.

38 "National Cyber Policy Framework for South Africa." 27.

39 "While the question of who has ultimate responsibility for cyber defence and operations may yet have to be settled, it can be argued that a major cyber-attack on a country has the potential to cause as much – if not more – damage than even a conventional attack. Thus there is a clear requirement for cyber defence to be part of the defence portfolio, and thus also for this to be addressed by the defence industry. That, in turn, creates the potential for close cooperation among all services, agencies and government departments engaged in this field." National Defence Industry Council. "Defence Industry Strategy", Version 5.8, (May 2017), Status: Draft: 31.

and commercial off the shelf products (COTS) at the expense of developing an in-depth and evolving domestic technical capacity. In this regard, it is vital that a balance is struck between producing "soft" cyber skills in the intelligence analysis sense and "hard" cyber skills in the technical arena.

This scenario calls for innovation and a departure from traditional defence recruitment methods. It is postulated that a diverse recruitment strategy should incorporate a drive to secure experienced ICT Security professionals alongside a programme to fast track the training of young graduates and school leavers who display a proven aptitude for cyber-related matters.[40] Such an approach would probably necessitate a combination of military and civilian personnel and possibly contractors working alongside one another. Inevitably, such structures are not favored; however, it is unlikely that the necessary expertise will be sourced without such a flexible approach to staffing.[41]

Lastly, the provision of cyber education and training will require well-defined learning pathways for the different streams. South Africa fortunately has a diverse tertiary education sector as well as tertiary military institutions of note. However, an approach which incorporates partnering with both will be required. Importantly, the distinction between training and education will also have to be made. As Esterhuyse points out:

> Military training is a continuous process that not only includes the learning or acquisition of initial skills, knowledge, attitudes and understanding necessary for the performance of tasks and roles, but also frequent rehearsals and practice.[42] Education on the other hand presents different competencies: the focus

---

40 "Defence Cyber Strategy" Netherlands Ministry of Defence, (2015) https://english.defensie.nl/topics/cyber-security/defence-cyber-strategy.

41 "Defence Cyber Strategy" Netherlands Ministry of Defence, (2015).

42 Abel Esterhuyse, "Professional Military Education in the South African National Defence Force.: The Role of the Military Academy." Unpublished Doctoral Dissertation, Stellenbosch University, (2007): 40.

> is on cognitive objectives written at the appropriate level of learning—knowledge, comprehension, application, analysis, synthesis, or evaluation—to develop the individual's ability to think. Education, thus, instils the mental flexibility to look beyond the horizon, to anticipate and to shape the future.[43]

Shaping these different pathways will be a core element in human resource development within the cyber corps. In addition to this, due to the different niche areas of expertise, a lifelong-learning pathway will be essential for all SANDF cyber personnel. An additional but imperative requirement is for a robust cyber awareness and resilience programme which covers all personnel throughout the security cluster.

In the quest to expand knowledge and capability, it is submitted that the government will need to establish and give credence to a triple helix type[44] of approach to cyber innovation. The need to partner with industry and academia in an effort to promote a coordinated national response to threats and skills shortages is pressing. There are, however, numerous challenges associated with this, one of which is the reluctance of higher education institutions to embrace what is viewed by society as a rather dark tradecraft. As Steven LaFountain, technical director of the National Security Agency in the US, commented, "Universities don't want to touch [hacking], they don't want to have the perception of teaching people how to subvert things."[45] Notwithstanding these misgivings, there is available expertise in the private sector which needs to be harnessed; likewise, the research ability within universities and the public research institutions should be expanded within structured and funded research programs in order to achieve and sustain a

---

43 Esterhuyse, "Professional Military Education in the South African National Defence Force: The Role of the Military Academy." 42.

44 The triple helix model of innovation refers to a set of interactions between academia, industry and governments, to foster economic and social development. See Ektowitz and Leydsdorf.

45 Newton Lee Laboratories, "Counterterrorism and Cyber Security - Total Information Awareness." Springer, New York, (2013): 144.

technological and scientific advantage. At this juncture, only two South African universities offer cyber-related qualifications—at a certificate level but not as a full degree.

To this end, knowledge exchange and collaboration with other countries would be a further important facet of such an approach. Finally, a concerted effort to narrow the civil-military gap, which is growing in South Africa, will be required. The DoD will need to work closely with elements of the private sector and research entities in order to address numerous cyber challenges and maintain fruitful partnerships. Cornish et al make the point that cyber tips the traditional balance of power where the state is the dominant entity:

> For its part, politics must also acknowledge the challenges of cyber warfare: its complexities must be extended back into the world of politics, questioning deeply embedded assumptions about the primacy of the state, the authority of government and the role of government agencies and the armed forces as providers of national security.[46]

It is of paramount importance that the South African State acknowledges the importance of public/private partnerships in combating the cyber threat. Aside from requiring expertise from the private sector, there are simply too many shared resources between the two entities: NCIs for one, but further to that, the entire communication infrastructure on which government operates is largely owned and operated by the private sector.

In summation, it is submitted that some of the proposed key strategic drivers for an overarching strategic framework are:

- The protection of South African national assets, citizenry, and economy.
- Foreign policy drivers with respect to South Africa's role on the African continent and the African security architecture.

46 Cornish, "On Cyber Warfare." vii.

- A collective approach to cyber defence within regional, economic, and historical alliances.
- Cyber power as an enhancement of state power.
- Reducing the civil-military gap in terms of cyber cooperation in South Africa.

In terms of determining more specific strategic considerations towards designing a militarily-orientated cyber warfare strategy, it is necessary to consider the current South African threat landscape insofar as it pertains to state-directed cyber threats.

Designing a Strategy for the South African Cyber War Threat Landscape

Greathouse offers some central elements in terms of strategic qualifiers and cyber warfare:

> Because cyber warfare is unconventional and asymmetric warfare, nations weak in conventional military power are also likely to invest in it as a way to offset conventional disadvantages. Going forward policy makers will be required to develop strategies which address the issues of cyber war. The difficulties of developing effective strategies will be compounded by a multitude of issues including: what qualifies as cyber war, should responses be the same as from attacks by state or non-state actors, does the state respond the same if elements of its civilian sector are attacked rather than the public sector, and whether an offensive or defense stance is necessary?[47]

From a threat perspective, it is commonly accepted that South Africa does not face an immediate conventional military threat. Defence planners, however, do draw attention to cyber threats:

47 Craig Greathouse, "Cyber War and Strategic Thought: Do the Classic Theorists still Matter?" in Cyberspace and International Relations, ed J.F. Kremer, J.F. and B. Müller, (2014): 22.

"Cyber and terror attacks remain a possibility to contemplate. Although no international armed conflict threat against South Africa is foreseen in the next five years."[48] Within the Fifth Domain of warfare, South Africa, like all states, should be anticipating cyber attacks from both state and non-state actors.

From an African security angle, South Africa, like the West, faces manifold challenges emanating from the so-called Arc of Instability which spans the torso of the African continent. These are primarily crime, illegal migration, and possible terrorist threats.[49] Migration brings with it economic and societal challenges, and the numerous outbreaks of xenophobic violence in South Africa over the past decade bear testimony to this. There are a number of large terrorist groups which are currently well established in the Arc, including Al Qaeda in the Islamic Maghreb (AQIM), El Shabaab, the Islamic State (ISIS), and Boko Haram.[50] These groups present a threat to the safety of the people of South Africa, and the influx of migrants from that region as a result of war and failed states provides a possible added dimension to this problem.

There are a number of aspects to this which require the attention of the authorities when creating both a cyber strategy and designing a cyber workforce. Social media undoubtedly provides a vehicle for the grooming and recruitment of future terrorists; one only has to look to ISIS and its metamorphosis into a so-called digital caliphate to understand the real and pervasive nature of this danger.[51] To this end, a significant drive to create both an advanced digital open source intelligence (OSCINT) as well as a social media intelligence (SOCINT) capability is required, regardless of the fact that Human Intelligence (HUMINT) remains the primary source of

---

48 DOD Annual Performance Plan, 8.

49 Solomon. "African Solutions to Africa's Problems?" 52-53.

50 Hussein Solomon "Jihad: A South African Perspective." SUN Media, Bloemfontein, (2013): 68.

51 Anne Speckhard et al, "Defeating ISIS on the Battle Ground as well as in the Online Battle Space: Considerations of the 'New Normal' and Available Online Weapons in the Struggle Ahead." *Journal of Strategic Security*; Vol. 9, ISS, 4, Winter (2016): 9.

reliable information within the African battle space. South Africa has already developed a significant and internationally-recognized capacity in OSCINT, so leveraging that platform in order to ensure the transition to enhanced digital capabilities will be essential.[52]

However, terrorism is also intertwined with cyber crime and provides both a vehicle for fund raising as well as money laundering. Cyber crime is potentially a massive issue due to the presence of global organized crime groups which have located their operations in South Africa. A number of scholars have repeatedly pointed to the issue of links between these criminals and terrorist groupings. The renowned terrorism scholar Hussein Solomon explains:

> There are a number of other reasons that make South Africa vulnerable to such Al-Qaeda and other Islamists' penetration. First, there are long borders and coastlines, which make the country increasingly porous. Second, this is made worse by the levels of bribery and corruption inside government departments facilitating ease of access into South Africa through fraudulently obtained passports and identity documents. Third, and closely linked to the latter is the presence of highly sophisticated criminal networks developing across southern Africa since the 1980s. Whilst organs of state are weak and corrupted, South Africa does not constitute a failed state as does Somalia—precisely the conditions under which such organised crime syndicates thrives as Mark Shaw explains, 'Organised crime operates best in the context of a corrupted state and organised business sector not one that has completely broken down. The existence of a relatively strong but penetrated state allows organised crime the luxury of using state institutions for profit, remaining relatively free from prosecution while continuing to operate in a comparatively stable environment.' These were to

52 NATO OSCINT Reader, NATO Europe, (February 2002): 95.

develop strong ties with radical Islamists who not only assisted them in terror financing but also in the penetration of organs of state.[53]

Obviously, this situation gives rise to the requirement for surveillance of possible suspects in this loop. This is an area fraught with difficulty on two levels. Firstly, there are the necessary but onerous requirements which the State must fulfil before the courts in terms of the Constitution before any type of electronic surveillance can take place. This is currently the subject of a heated national debate as the State Security Agency is in the process of trying to move the new Cyber Security Bill through parliament. Civil society organizations, while for the most part recognizing the State's need to expand cyber measures, are naturally concerned about the possible abuse of such power, primarily with regards to citizen and political surveillance. However, it is undoubtedly also true that both criminals and terrorists make substantial use of the electronic spectrum to communicate, recruit, manage their finances, purchase weapons, and plan.[54]

The technical and constitutional challenges in this domain are of a global nature and pose a problem to most governments. Take for example the utilization of highly-encrypted messaging services such as Whatsapp and Telegraph which are decidedly favored by terrorist cells when planning and executing attacks. The British government has already displayed its dismay and annoyance while trying to find a solution to this issue and meeting with seeming intransigence from tech companies.[55] This points to the asymmetrical nature of cyber warfare where lone individuals can mount serious challenges to security with even limited digital knowledge and equipment. To this end, an advanced technical and

53 Hussein Solomon, "Jihad: A South African Perspective." 6.

54 Cornish, "On Cyber Warfare." 8.

55 Parmy Olson, "U.K. Calls For Backdoor To WhatsApp After London Attacks." Forbes, (27 March 2017). https://www.forbes.com/sites/parmyolson/2017/03/27/britain-backdoor-whatsapp-london-attacks/#74073le11a23

intelligence capability will need to be developed which functions within a well-defined policy and legislative environment.

Another type of non-state actor, the hacktivist community, form a different yet rather persistent threat to many African governments. Between 2010 and 2016, hacktivist attacks constituted 31% of all cyber-related attacks in South Africa and, for the most part, these were aimed either at the State or at parastatal corporates. Van Niekerk attributes this to "a growing protest and revenge dimension in South Africa's cybersecurity risk profile."[56] What is of some concern is that a number of these parastatals such as ESKOM (the state-owned electricity supplier) represent key positions within the national critical infrastructure framework.[57]

Espionage, the threat to Critical Information networks, and NCI form the basis of additional immediate cyber defence threat targets. It is probably fair to postulate that most states can and do indulge in espionage.[58] Cyber espionage provides the perfect vehicle for such activities, particularly where certain states possess an advanced degree of cyber capability. It is in this area where the SANDF needs to be particularly vigilant as espionage activity is probably a given, and, secondly, the harsh reality is that oftentimes networks are breached and it can be months before such a breach is detected, if ever. The same holds true in terms of the protection of networks in ensuring that information is confidential, secure, always available, and easily replicated. As the DoD is reasonably reliant on its digital information platform, planning and mitigation strategies in this area are paramount.

Further to this is the need to protect South Africa's defence industry, which is a sizeable export industry. Responsibility for armaments procurement, development, and manufacturing rests with the state-owned enterprise ARMSCOR, which works closely with the DoD. ARMSCOR has already fallen prey to a rather

56 Van Niekerk, "An Analysis of Cyber-Incidents in South Africa." 123.

57 Van Niekerk, "An Analysis of Cyber-Incidents in South Africa." 127.

58 Ibid.

public hack by Anonymous Africa in 2016, which was achieved using a relatively simple SQL injection method which compromised the details of its procurement system.[59] Of more concern is the protection of ARMSCOR's intellectual property as it is a global leader in certain niche technologies. Another vital consideration is securing and assessing the software on weapons systems developed for the SANDF, particularly where this has been developed by non-South African companies. Such enterprises will require ongoing efforts and vigilance.

The draft Defence Industry Strategy of 2017[60] alludes to the growing importance of technology in future product developments in support of the Defence Strategic Trajectory. Weapons systems will increasingly rely on secure technologies. This has software security implications both during the developmental phase as well as the operational phase. Having said this, it is also vital to ensure that SANDF forces can operate independently of networked digital technologies within the African battle space due to localized constraints, and this will be a key element of the cyber strategy.

The issue of organizational structure within the SANDF also requires attention, as it speaks to the question of inter-operability. More specifically, this refers to the question of where to place an entity such as a cyber command within the force structure. There is no common cause amongst armed forces around the world on this matter and cyber forces can be found within signals formations, intelligence formations, or even as stand-alone high-level commands. The SANDF could consider situating its cyber forces within its technical arm (the Command Management Information Formation), within its Intelligence structures, or possibly within the Joint Operations Environment. It is contended that joint and inter-operability with all arms of the SANDF is a

59 DefenceWeb, "Armscor Website Hacked." (13 July 2016) http://www.defenceweb.co.za/index.php?option=com_content&view=article&id=44258:armscor-website-hacked&catid=90:science-a-defence-technology&Itemid=204.

60 National Defence Industry Council. Defence Industry Strategy, Version 5.8, May 2017, 106.

key requirement and, therefore, the placement of cyber command is of critical import.[61]

Drawing on the preceding discussion, certain key requirements and outcomes towards building a cyber warfare strategy for South Africa are proposed below:

SA Cyber Warfare Strategy: Requirements

- Coordinated governmental response and planning, particularly within the Security Cluster.
- An overarching national cyber strategy and policy.
- Resources—financial/technical/human.
- Defence partnerships at inter-state level.
- A triple-helix approach to cyber defence.
- Alignment of cyber warfare strategy and doctrine with conventional South African military doctrine.

SA Cyber Warfare Strategy: Proposed Outcomes

- A comprehensive cyber warfare strategy within a robust policy environment.
- Annual strategic review mechanism with metric enablers.
- A primarily defensive cyber posture.
- Cyber doctrine development which is suitable for the African battle space.
- Cyber Resilience within the entire Joint Security Cluster.
- A regional and collective focus on cyber defence, particularly within SADC.

In closing, a significant point is that in the determination of strategy and policy, it is essential that defence planners consistently revisit both, given the constantly-evolving technological landscape and the changing capacity of threat actors. It would be remiss to discuss matters of strategy without briefly touching on the

61 Cornish, "On Cyber Warfare." 13.

importance of metrics and review. Given the fact that cyber strategy is in its infancy as both a scholarly and military field, coupled with persistent and rapid advances in technology, strategists and planners in this domain are on a constant learning curve. Strategy is by its nature an iterative process and it is rather crucial that cyber incidents, responses, and initiatives are well documented. A system of metrics which measure incidents, successes, and failures is imperative to the ongoing development of both strategic and tactical guidelines. Likewise, the cyber warfare strategy should be reviewed and updated regularly, with the assistance of such metrics, in order to remain relevant in meeting a continuously changing range of threats and actors. Greathouse points to three important areas which require ongoing revision: "what are viable targets, second how to deal with non-state actors, finally what offensive/defensive balance will be pursued?"[62]

Conclusion

A well-articulated and responsive cyber warfare strategy forms a fundamental foundation towards responding to the persistent threat of cyber attacks faced by nation states in the present era. South Africa, like many other countries, is grappling with numerous challenges relating to cyber threats. It is incumbent on the State to create a strong framework in order to counter and adequately respond to cyber attacks and incursions. This entails a number of aspects, from ensuring a comprehensive legal framework, to working closely with other states through to the formulation of a national cyber strategy and a cyber warfare strategy. The need for the State to rapidly develop capacity, establish cyber resilience within the population, and deploy properly-functioning cyber response entities and to develop an effective reporting system is further central to ensuring progress.[63] There is a pressing necessity

62 Greathouse, "Cyber War and Strategic Thought: Do the Classic Theorists still Matter?" 38.

63 Van Niekerk, "An Analysis of Cyber-Incidents in South Africa." 128.

to engage and partner more closely with both the private and education sectors in order to strengthen human capital growth and facilitate seamless responses to threats to the economy, the populace, and the State. The effective definition and implementation of a cyber warfare strategy cannot take place without the above measures having been instituted.

A cyber warfare strategy will require careful planning and the input of expertise drawn from the military, academia, and the private sector. This presents something of a unique situation to military planners in most countries. However, cyber warfare challenges traditional military norms and requires enormous flexibility of thought and therefore a departure from traditional conventions. For the SANDF this will mean a new approach to recruitment, operational security, resource allocation, and force design. It will also entail playing a leadership role in developing cyber capacity within SADC. There are a number of strengths which South Africa has in its favor which can be leveraged to this end: an advanced capability within the information technology sector, the most developed science system in Africa, and a significant armaments industry coupled with a military research and development capacity. The development of the imminent cyber warfare strategy must draw on these strengths.

Bibliography

Beresford, Alexander. "A responsibility to protect Africa from the West? South Africa and the NATO intervention in Libya." *International Politics—A Journal of transnational issues and global problems*. Vol 52 (93), (2015) 288-304.

Cornish, Paul, David Livingstone, Dave Clemente and Claire Yorke, C. "On Cyber Warfare." *A Chatham House Report* (November 2010) 1-38.

DefenceWeb, "Armscor Website Hacked." 13 July 2016. http://www.defenceweb.co.za/index.php?option=com_content&view=article&id=44258:armscor-website-hacked&catid=90:science-a-defence-technology&Itemid=204

Department of Defence and Military Veterans. "*DOD Annual Performance Plan 2016*" (2016).

Department of Defence and Military Veterans. "*DOD Planning Instruments for 2015 to 2020,*" (2015).

Department of Defence and Military Veterans. "*South African Defence Review 2015.*" South African Department of Defence, (2015).

Department of Justice and Correctional Services. "*Cybercrimes and Cybersecurity Bill.*" As introduced in the National Assembly (proposed section 75), (2017) Republic of South Africa.

Esterhuyse, Abel. "Professional Military Education in the South African National Defence Force.: The Role of the Military Academy." Unpublished Doctoral Dissertation, Stellenbosch University, (2007): 1-340.

Esterhuyse, Abel. "The South African Threat Agenda: Between Political Agendas, Perceptions and Contradictions." *S&F Sicherheit und Frieden*, (2016) (Seite): 191 – 197.

Goldstuck, Arthur. "Massive SA growth of 4G, Wi-Fi, in next five years." *Mail and Guardian*, 22 Apr 2015. https://mg.co.za/article/2015-04-22-massive-sa-growth-of-4g-wi-fi-in-next-five-years

Greathouse, Craig B. "Cyber War and Strategic Thought: Do the Classic Theorists still Matter?" in Cyberspace and International Relations, ed J.F. Kremer, J.F. and B. Müller, Springer-Verlag, Berlin Heidelberg (2014): 21-40.

Iasiello, Emilio. "Is Cyber Deterrence an Illusory Course of Action?" *Journal of Strategic Security*, 7, no. 1 (2013): 52-67.

Joint Communiqué on the occasion of the 11th Meeting of the United Kingdom – South Africa Bilateral Ministerial Forum, London, 19 October 2015. http://southafricahouseuk.com/documents/11bilatrlfrm.pdf

Jones, Andrew and Kovacich, Gerald L. "Global Information Warfare: The New Digital Battlefield." (Second Edition, Taylor and Francis, USA), (2016).

Manson, G.P. "Cyberwar: The United States and China Prepare For the Next Generation of Conflict." *Journal Comparative Strategy*, Volume 30, 2, (2011): 121-133.

Mbanaso, Uche M. "Cyber warfare: African research must address emerging reality." *The African Journal of Information and Communication (AJIC)*, (18) (2016) 157-164.

National Defence Industry Council. "Defence Industry Strategy", Version 5.8, (May 2017), Status: Draft.

NATO Open Source Intelligence Reader, NATO Europe, (February 2002): 1-110.

Neethling, Theo J. "South Africa's Foreign Policy and the BRICS Formation: Reflections on the Quest for the 'Right' Economic-diplomatic Strategy." *Insight on Africa*, Vol 9 (2017): 39-61.

Neethling, Theo J. "South Africa and AFRICOM: Reflections on a lukewarm relationship." *South African Journal of International Affairs*. Vol 22: 1,(2015): 111-129.

Netherlands Ministry of Defence, "Defence Cyber Strategy" (2015) https://english.defensie.nl/topics/cyber-security/defence-cyber-strategy

Newton Lee Laboratories, "Counterterrorism and Cyber Security - Total Information Awareness." Springer, New York, (2013): 1-235.

Nye, Joseph. "Cyber Power." Monograph, Harvard University Belfer Center for Science and International Affairs. May, (2010): 1-30.

Olson, Parmy. "U.K. Calls For Backdoor To WhatsApp After London Attacks." Forbes, 27 March 2017. https://www.forbes.com/sites/parmyolson/2017/03/27/britain-backdoor-whatsapp-london-attacks/#740731e11a23

Robinson, Michael., Kevin Jones. and Helge Janicke. "An Introduction to Cyber Peacekeeping." https://arxiv.org/pdf/1710.09616.pdf, (Oct 2017): 18-22.

Savides, Mathew. "Fewer protests in 2016, but they were more violent." https://www.timeslive.co.za/news/south-africa/2017-02-01-fewer-protests-in-2016-but-they-were-more-violent/

Sigholm, Johan. "Non-State Actors in Cyberspace Operations." *Journal of Military Studies*, 4(1) (2016), 1-37.

Solomon, Hussein. "African Solutions to Africa's Problems? African Approaches to Peace, Security and Stability." *Scientia Militaria - South African Journal of Military Studies*, V 43, N 1, (May. 2015): 45-76.

Solomon, Hussein. "Jihad: A South African Perspective." SUN Media, Bloemfontein, (2013): 1-107.

Speckhard, Anne; Ardian Shajkovci, and Yayla, Ahmet, S. "Defeating ISIS on the Battle Ground as well as in the Online Battle Space: Considerations of the 'New Normal' and Available Online Weapons in the Struggle Ahead." *Journal of Strategic Security*. Vol. 9, ISS, 4, Winter (2016): 1-10.

Stand-To. The Official Focus of the US Army. "Shared Accord-2017". https://www.army.mil/standto/2017-07-17

State Security Agency. "*National Cyber Policy Framework for South Africa*." South African Government Gazette No. 39475 of 4 December (2015).

Tass Russian News Agency. "Russian, South African top diplomats ink cyber security cooperation deal." 4 September 2017, http://tass.com/politics/963564

Tatar, Ünal; Çalik Orhan; Çelik Minhac and Karabacak Bilge. "A Comparative Analysis of the National Cyber Security Strategies of Leading Nations." *International Conference on Cyber Warfare and Security*. Reading: 211-X. Reading: Academic Conferences International Limited. (2014).

United Nations. "*Global Cybersecurity Index (GCI).*" International Telecommunications Union, (2017).

Van Niekerk, Brett. "An Analysis of Cyber-Incidents in South Africa." *The African Journal of Information and Communication (AIJC)*, (20): 115-128.

Vines, Alex and Elizabeth Sidiropoulos. "UK and South Africa: A Relationship Worth Maintaining." Chatham House, 2013, https://www.chathamhouse.org/media/comment/view/194212.

Noëlle van der Waag-Cowling
Department of Military Strategy, Faculty of Military Science, Stellenbosch University and Associate Researcher, Centre for Conflict, Rule of Law and Society, Bournemouth University.
Private Bag X2, Saldanha, South Africa, 7395. noelle@ma2.sun.ac.za

3

Hybrid Wars: The 21st Century's New Threats to Global Peace and Security

Sascha-Dominik Dov Bachmann, Bournemouth University
Håkan Gunneriusson, Swedish Defence University

This article was originally published in *Scientia Militaria, South African Journal of Military Studies*, Vol 43, No. 1, 2015, pp. 77 – 98. It is reprinted with permission.

Abstract

This article discusses a new form of war, 'hybrid war,' with inclusion of aspects of 'cyber-terrorism' and 'cyber-war' against the backdrop of Russia's 'Ukrainian Spring' and the continuing threat posed by radical Islamist groups in Africa and the Middle East. It also discusses the findings of an on-going hybrid threat project by the Swedish Defence College. This interdisciplinary article predicts that military doctrines, traditional approaches to war and peace and their perceptions will have to change in the future.

Introduction

The so-called 'Jasmine Revolution' during the Arab Spring of 2011 challenged the political order in the Maghreb and the whole Middle East. While some of the protests led to actual regime changes and a move towards freedom and democracy—such as in Tunisia—events in other states in the region, such as Bahrain and Syria, had been less successful and saw the return of the 'old order' of autocratic governments. The collapse of Muammar Gaddafi's regime in Libya, the on-going civil unrest in Egypt between supporters of the ousted hard-line Muslim brotherhood and the military government, the on-going brutal Syrian conflict and the collapse of Iraq after the withdrawal of the USA have all significantly contributed to the proliferation and the ascent of evermore powerful and murderous terrorist groups and organizations across the region.

The use of 'cyber'[1] and kinetic responses to international terrorism have increasingly blurred the traditional distinction between war and peace. Such a distinction was replaced by the recognition of a notion of new, multi-modal threats, which have little in common with past examples of interstate aggression. These new threats to global peace and security seriously threaten our modern Western way of life within the context of the present 'steady-state' environment at home (and against the backdrop of the ongoing asymmetric conflicts in Afghanistan, Pakistan, Mali, Somalia, Kenya and Yemen). These new wars "along asymmetric lines of conflict"[2] constitute "a dichotomous choice between counterinsurgency and conventional war"[3] and challenge traditional concepts of war and peace.

This article[4] firstly reflects on the new notion of so-called 'hybrid threats' as a rather new threat definition and its (temporary) inclusion in the North Atlantic Treaty Organization's (NATO) new comprehensive defense approach with a reflection on the last Swedish experiment. Secondly, it discusses the use of 'cyber' in the context of 'hybrid threats' before it, thirdly, addresses some implications for military doctrine arising from such threats. The article concludes with a brief outlook on new dimensions of possible future threats to peace and security by highlighting the evolvement of the concept of 'hybrid threats' into 'hybrid war' by reflecting on security issues arising.

'Hybrid threats' as challenges to peace and security

The novel concept of hybrid threats first gained recognition when Hezbollah had some tangible military success against the Israeli Defense Forces (IDF) in Lebanon 2006 during the Second Lebanon War.[5] Ironically, the definition of 'hybrid' then was that a non-state actor showed military capabilities one originally only associated with state actors.[6] Multimodal, low-intensity, kinetic as well as non-kinetic threats to international peace and security include cyber war, asymmetric conflict scenarios, global terrorism, piracy, transnational organized crime, demographic challenges, resources security, retrenchment from globalization, and the proliferation of weapons of mass destruction. Such (multi-)modal threats have become known as 'hybrid threats.'[7] Recognized in NATO's Bi-Strategic Command Capstone Concept of 2010, hybrid threats are defined as "those posed by adversaries, with the ability to simultaneously employ conventional and non-conventional means adaptively in pursuit of their objectives."[8] Having identified these threats, NATO undertook work on a comprehensive conceptual framework, as a Capstone Concept, which was to provide a legal framework for identifying and categorizing such threats within the wider frame of possible multi-stakeholder responses. In 2011, NATO's Allied Command Transformation (ACT), supported by the US Joint Forces Command Joint Irregular Warfare Centre (USJFCOM JIWC) and the US National Defence University (NDU), conducted specialized workshops related to Assessing Emerging Security Challenges in the Globalised Environment (Countering Hybrid Threats [CHT]) Experiment.'[9] These workshops took place in Brussels (Belgium) and Tallinn (Estonia) and were aimed at identifying possible threats and at discussing some key implications when countering such risks and challenges. In essence, hybrid threats faced by NATO and its non-military partners require a comprehensive approach allowing a wide spectrum of responses, kinetic and non-kinetic, by military and non-military actors. In a 2011 report, NATO describes such threats as,

> Admittedly, hybrid threat is an umbrella term, encompassing a wide variety of existing adverse circumstances and actions, such as terrorism, migration, piracy, corruption, ethnic conflict, etc. What is new, however, is the possibility of NATO facing the adaptive and systematic use of such means singularly and in combination by adversaries in pursuit of long-term political objectives, as opposed to their more random occurrence, driven by coincidental factors.[10]

The same report underlines that hybrid threats –

> ... are not exclusively a tool of asymmetric or non-state actors, but can be applied by state and non-state actors alike. Their principal attraction from the point of view of a state actor is that they can be largely non-attributable, and therefore applied in situations where more overt action is ruled out for any number of reasons.

The findings of the two workshops were published in the ACT's final report and recommendations in 2011. However, due to a lack of financial resources in general and an absence of the political will to create the necessary 'smart defense' capabilities among its member states, NATO decided in June 2012 to cease work on CHT at its organizational level while encouraging its member states and associated NATO Excellence Centres to continue working on hybrid threats.

In 2012, the Swedish National Defence College as a Partnership for Peace (PfP) partner[11] conducted its own hybrid threat experiment.[12] The scenario dealt with a fictitious adversary in the East, not very dissimilar to Belorussia, except that it was an island kingdom in the Baltic Sea. The situation deteriorated to the point where neighboring states were directly affected by a mix of conventional military and hybrid threats. More traditional threats arose from the attempt to sink a hijacked oil tanker in the middle of the sensitive maritime environment zone, launching a small group of Special Forces operatives (SFOs) in Swedish territory and hiring Somali pirates to hijack Swedish vessels off the Horn of Africa. The latter showed how a conflict could spread from being very local in one part of the world to involve remote hotspots in Africa. In this case, the problems at the Horn of Africa could legitimize actions and events, which originally had their roots in Northern Europe. The participants of the experiment acted as a committee of advisers for the Swedish government, and their individual roles represented their normal functions: from members of the armed forces and national support agencies to the university sphere, the pharmacological industry, banking and internet security. The experiment showed that existing and established standard operation procedures (SOPs) made responding to specific threats rather efficient. This was mostly due to already established command and control as well as communication and coordination assets and abilities. The experiment did however also show the existence of shortcomings when countering multi-modal threats due to the absence of a nationally defined comprehensive approach for a joint interagency approach. With SOPs in place and lacking a uniform command and control structure, it can also become harder to respond in a tailored and united way for government agencies, as all contributing agencies have their respective tasks and procedures. This lack of comprehensive joint action and coordination is highlighted by the fact that the government in the scenario did not have the authority to direct and control the work of subordinate but autonomous agencies.[13] The participants of the hybrid threat experiment did recognize that a coming hybrid conflict would lead to new levels of threat and response complexity and that there was a need for active, uniform, and collective leadership beyond SOPs.[14] The participants identified as a weakness the lack of a comprehensive response and coordination between agencies such as the armed forces, the civil defense assets and other civilian actors, such as IT specialists and pharmaceutical experts.[15] With a shrinking defense budget, the downscaling of agencies and an obvious lack of civil society to accept the potential existence of such threat in the future, it seems unlikely that these shortcomings will be addressed in the near future.

In an African and Middle Eastern context, one cannot generalize as these states differ in terms of stability and strength regarding the capacities of their security assets. A state such as South Africa should and could rely very much on SOPs in order to have a constant high readiness against unsuspected threats. Other countries with weaker infrastructures and resources cannot expect their agencies to react swiftly when faced with ad hoc security challenges. The recommendation should then be to have very able actors (rather than structures, which the SOP demands) at key positions (at ministerial level and the level below) who can understand the threat and swiftly tailor a suitable response with the resources the state has at hand itself and with allied states. The latter is important in general and certainly so in Africa. As the borders have a colonial past, one should expect hybrid threats stemming from non-state actors (NSAs), which will eventually encompass a number of states.

Worrying—and of particular relevance in the context of hybrid threats—is the danger of proliferation of advanced weapon systems by NSAs associated with radical Islam, as for example the Islamic State of Iraq and al-Sham (ISIS) in Syria and Iraq as well as the increasing use of new technologies by NSAs. The last Israel–Gaza conflict highlights these developments: new technologically advanced rocket systems, supplied by Iran to their terrorist proxy Hamas, were used against Israel. The capability of the Fajr (Dawn) 5 rocket to reach both Tel Aviv and Jerusalem has been shown and has once more shown the vulnerability of Israel as a state when it comes to conventional, kinetic threats.

Against the backdrop of the on-going conflict in Ukraine and the classification of the conflict as a 'hybrid war' by Ukraine's national security chief,[16] NATO's decision to discontinue working on the hybrid concept as an organizational objective might turn out to have been made too early.[17]

The role of 'cyber-space' in hybrid threat scenarios in post-Cold War security

Despite NATO's failure to agree to a joint and comprehensive approach in countering hybrid threats, there is little doubt that "hybrid threats are here to stay."[18] Even a mainly conventional war will have a 'hybrid' element such as for example a 'cyber-attack,' 'bio-hacking,' and even 'nano-applications.'[19] Old threats, such as nuclear threats, can these days be reconsidered as within reach for state actors. Warnings have already been made that some university courses in nuclear technology might be in danger of being used by terrorist organizations.[20] Future attackers will rely increasingly on technological and scientific ways to execute their operations, and one of the documented examples is the use of 'cyber-space' for carrying out or controlling 'hybrid threats.'

'Cyber-conflict' and 'cyber-war' serve as examples of the use of new technologies within the scope of hybrid threats. Cyber-war[21] basically refers to a sustained computer-based cyber-attack by a state (or NSA) against the IT infrastructure of a target state. An example of such hostile action taking place in the fifth dimension of warfare is the 2007 Russian attempt to virtually block out Estonia's internet infrastructure as a unilateral countermeasure and retribution for Estonia's removal of a WWII Soviet War Memorial from the center of Tallinn.[22] Governmental and party websites as well as businesses were severely obstructed by this incident of cyber warfare, when Russian military operations were augmented by cyber operations against Georgia. This incident was followed by the employment of cyber measures in connection with the Russian military campaign in Georgia in 2008. Russia once again acted in a way which utilized the potential of the hybrid threat as a military strategy and modus operandi, this time in the Crimea.

Another example of how multi-modal threats, asymmetric terror and warfare are supplemented by terrorist (dis)information campaigns can be seen in the Israel–Gaza conflict. Then and now, Hamas has been employing tools and strategies of disinformation normally associated with clandestine psychological operations (PsyOps) of traditional military state actors, such as the sending of emails and text messages with hoax news updates as well as propaganda 'news flashes' sent to Israeli and non-Israeli email addresses and cell phones and the use of the internet to disseminate their propaganda.[23] During the eight days of conflict, text messages were sent which warned, "Gaza will turn into the graveyard of your soldiers and Tel Aviv will become a fireball."[24]

The (reported) use of a sophisticated computer worm to sabotage Iran's nuclear weapons programs, called Stuxnet, by presumably Israel, has highlighted both the technical advancement, possibilities as well as potential of such new means of conducting hostile actions in the fifth dimension of warfare.[25] The continuing and intensifying employment of such cyber-attacks by China against the USA, NATO, the European Union, and the rest of the world has led the USA to respond by establishing a central Cyber War Command, the United States Cyber Command (USCYBERCOM) in 2010[26] to "conduct full-spectrum military cyberspace operations in order to enable actions in all domains, ensure US/Allied freedom of action in cyberspace and deny the same to their adversaries."[27] Following these developments—and perhaps supplementing the work of USCYBERCOM—NATO set up a special hybrid threat study group, which is studying possible responses to such threats, the so-called NATO Transnet Network on Countering Hybrid Threats (CHT).[28] 'Cyber' in the context of armed conflict does not necessarily establish genuinely new categories of conflict per se; it rather constitutes another and improved 'tool' of warfare, namely 'cyber warfare.' The military will find new ways to conduct its operations by militarizing 'cyber-space' as a force multiplier and operational capability enhancer, and will continue to operate at the tactical, operational, or strategic level. The increasing hostile use of 'cyber-space' by NSAs to further their economic, political, and other interests, and the present problem of clear accreditation of the originators of cyber activities make it increasingly hard to identify and counter such threats. Terrorist NSAs (or terrorist proxies of a state sponsor such as Iran and Syria) are increasingly using cyber capabilities in the wider sense to augment their attack capabilities. Apart from the above-mentioned use of 'cyber-space' by Hamas as a means of disinformation during the last Israel–Gaza conflict, ISIS (Islamic State in Iraq and the Levant) has been successful in utilizing the 'cyber-space' for self-promotion and as a means of psychological warfare in its operations in Iraq and Syria.

One such example of the role of the internet and social media as an enhancer and force multiplier for terrorist activities can be found in the Mumbai attacks in India in 2008. Terrorists from Pakistan attacked the city, with a particular focus on the Taj Mahal Hotel.[29] Tactical intelligence during the raid was gathered from social media and the exploitation of existing mass media such as cable TV. Readily available home electronic equipment and cell phones were used as means of 'command and control.' Terrorist operatives on the ground were directed by their handlers in what can only be described as a classic war (situations) room in Pakistan. They were in permanent cell phone contact with the field operators in Mumbai, and were able to use both internet and major television channels for a situation update on the evolving situation on the ground, comparable to a situation report (SITREP) used by conventional armed forces. Live coverage of the attacks was made available by news channels, and as a novelty, by the social media, such as Flickr, Twitter, and Facebook. The handlers of the operation 'data mined' and compiled this information in real time and communicated operation-relevant information directly to the terrorists through the use of smartphones.[30] What one could observe in the Mumbai example was the amazing readiness, availability, and affordability of using new technologies for setting up an effective and workable system of 'command and control.'

This observation is a post-Cold War reality and a direct result of globalisation and technical advancement. The ways of accessing information in cyberspace are changing rapidly and are becoming increasingly hard to counter. One recent example of an ingenious way of 'hacking' into otherwise protected sources involved the use of Google programs for inserting a so-called 'backdoor' Trojan for the purpose of data theft later.[31] Using the Google server, which already had access to the information of interest, hackers bypassed any firewall used by the 'target.' Another example of using an otherwise 'innocent' host like Google for carrying out 'cyber attacks' took place in late 2012 when hacker 'vandals' defaced Pakistan's Google domain along with other official Pakistan websites.[32] Other examples are the use of Thingbots—such as TVs, media players, routers, and even a refrigerator—to send out spam in a coordinated and prolonged fashion.[33] While spam is mostly used for phishing activities, it also could be used for DDoS attacks (distributed denial of service attacks).

To summarize, one could state that the combination of new technology and the availability of these 'cyber'-supported or 'cyber'-led hybrid threats is what make these threats so potent. Command and control capabilities can be established in a relatively short time and without too much effort, and the media could be used for influencing the public opinion as a means of 'PsyOps,' both at home and abroad. 'Cyber threats' in general strike at the core of modern warfighting by affecting command and control abilities, which have become vulnerable to such 'cyber attacks.'

Hybrid Threats and Military Doctrines

Military doctrines provide guidance for the military logic of operational practice. It is therefore alarming that most Western military doctrines are apparently unprepared when it comes to hybrid threats. It seems as if NATO's inability and perhaps unwillingness to formulate a binding comprehensive NATO approach to hybrid threats is a testament to the perseverance of an overwhelmingly conservative military doctrinal approach. Time will tell whether this is to change. Latvia regards the 2014 events in Ukraine as clear evidence that NATO is unwilling and unable to provide protection at all if Russia was to repeat its Crimean operation in the Baltic States. A suggestion was made to change NATO's Washington Treaty so that Article 5 can deal with this kind of hybrid threats.[34] This is of course very unlikely as few of the NATO member states have anything at all to gain from military confrontation with Russia. It does however send a message to all states within the Russian interest sphere that there should be no doubt about Russia's strength and, correspondingly, NATO's weakness in this part of the world.

The failure of defining a NATO policy on countering hybrid threats is even more unfortunate given that the USA has a national military security strategy in place, which recognizes certain hybrid threats as part of new and existing threats to its national security.[35]

This failure may have its cause in a continuing Cold War-rooted psychology and thought among the political actors. During the Cold War, the world was locked in an intellectual doctrinal approach which viewed all conflicts in the context of the global ideological struggle coded by the laws and political paradigm of its time. Once the Cold War had come to an end in 1991, new national conflicts arose along once pacified conflict lines. This new era manifested itself in, for example, the bloody conflicts in the Balkans in the 1990s as a consequence of the breakup of the old communist regime, and the various conflicts on the territory of the former Soviet Union. While the Cold War was not necessarily only about the conflict between two opposing superpowers, nor exclusively about ideological confrontation, it nevertheless led to a strict division of the world and its conflicts into two major ideological spheres with only few exceptions, namely the spheres of the US-led West versus the Soviet-led East. This division made potential threats more foreseeable and even 'manageable.'

Since the end of the so-called 'Cold War,' the world has changed dramatically and it is clear that this is also affecting military operations and doctrines. While the collapse of the Soviet Union and the Warsaw Pact in 1991 removed the original *raison d'être* of the Alliance, the prospect of having to repel a Soviet-led attack by the Warsaw Pact on Western Europe, the end of the Cold War also ended the existing balance of power after World War II and led to a 'proliferation' of armed conflicts around the globe. It seems as if the use of inter- state force has once more become 'acceptable,'[36] as highlighted in the two 'War on Terrorism' campaigns, the Russian–Georgian conflict of the summer of 2008, the NATO-led Libyan Intervention of 2011, and Russia's recent operations in the Crimea and Ukraine proper. This potential for future interstate conflict adds to the above-discussed proliferation of 'hybrid conflict' where non-state actors have become very successful actors, aggressors respectively, in an inter- and intrastate conflict setting.

The end of the Cold War gave rise to a new way of thinking, which was no longer based solely on technological capabilities and/or sheer numerical superiority. It is possible to view the European postmodernism and the 'fourth generation warfare' following 9/11 as parallel tracks, with the latter challenging the paradigm of the

Western positivistic materialism.[37] While military academics in the Western world do not lack warnings about the new challenges brought by these changes, it will eventually be up to politicians to 'drive' new initiatives, a prospect often marred by 'Realpolitik,' which will determine any policy in the end.

How does that affect military (and) security doctrines? Doctrinal changes for the military will depend on how the laws of war and the use of force will be shaped and this, in turn, will be shaped by the practice of those who should adhere to it. This has been highlighted by examples where legitimacy has been ignored on behalf of Realpolitik, as the operations in Afghanistan and Iraq show. What one can hope for in military doctrine is an integration of the rest of society in the common effort to protect itself from all forms of threats, conventional interstate aggression as well as new hybrid threats. One such example is the recent suggestion by the UN that states should and ought to be more proactive when it comes to fighting the use of the internet by terrorists.[38] Only society as a whole can protect itself, a task which is not limited to the military only, but which, on the other hand, cannot take on this huge task alone. An integration of the capabilities at interstate level, something NATO refers to as 'smart defense,' and increased defense cooperation may be the only way forward to counter the multitude of ever-evolving threats in the future.

The capacity of NSAs to copy the command and control structures of conventional military has increased with the ready availability of mass-produced information technology and the possibility to tap into open sources for 'data mining.' These developments have changed the traditional view of asymmetric warfare, where an AK-47 and the insurgent's morale were traditionally the only and often most important factors in achieving victory. The asymmetric warfare concept used to be an idiom to describe war against opponents who also used to be weaker in terms of available weaponry and utilization of technology.

Hybrid threats as such are not new threats; what is new is the recognition that such multi-modal threats command a 'holistic' approach, which combines traditional and non-traditional responses by state and NSAs as well, such as multinational companies. Responses to hybrid threats have to be proportionate and measured: from civil defense and police responses to counterinsurgency (COIN) and military measures. On the other hand, even NATO has something to win on the lack of codification. There will be a grey area of conflict where all actors can act—states, as well as non-states. Not that this seems to be the recipe for a bright future, but at least the possibilities for action will be there, even for Western states.

Hybrid threats and their possible responses challenge Carl von Clausewitz's dogma of war as constituting "a mere continuation of [state] politics by other means"[39] and might degrade the definition by Clausewitz into an early modern/modern *moyenne durée* definition of armed conflicts to use Fernand Braudel's term,[40] namely that of a permanent state of war and conflict of varying intensity. NATO followed this rationale in its approach to countering hybrid threats, as they wanted a conventional threat element in the hybrid threat definition in order to ensure the operational usefulness of its conceptual approach. NATO's failure to formulate a comprehensive response strategy to asymmetric and 'hybrid' threats is an omission which will come at a cost in the future. International cooperation on capabilities is the *sine qua non* of future counter-strategies in order to respond to such threats and to be prepared for evolving new threats. This necessity of being prepared reflects on Sun-Tzu when he said, "Victorious warriors win first and then go to war, while defeated warriors go to war first and then seek to win."[41]

Conclusion: From Hybrid Threats to Hybrid War

Russia and Ukraine

Russia's offensive policy of territorial annexation (of the Crimea), the threat of using military force and the actual support of separatist groups in the Ukraine have left the West and NATO practically helpless to respond.[42] NATO seems unwilling to agree on a more robust response; thus, revealing a political division among its member states. This unwillingness can partly be explained by Europe's dependency on Russian gas supplies but also by the recognition of legal limitations and considerations, such as NATO's Article 5 (which only authorises the use of collective self-defence in cases of an attack on a NATO member state). It does not seem far-fetched to see the events of spring 2014 as the emergence of a new power balance in the region. As it was the case with the two historical examples, the overall outcome will be different from what was initially expected. The recent events have also brought Russia back into the region as the main player. Russia's (re-)annexation of the Crimea in April 2014 is a fait accompli and unlikely to be revised anytime, and the on-going support of separatist groups in the eastern parts of the Ukraine where the Russian-speaking minority is in the majority, such as Donetsk and Luhansk, has seen an increase in open military combat.[43] Ukraine is already a divided country, with fighting taking place along its ethnic lines. The break-up of the old Yugoslavia in the 1990s and its ensuing humanitarian catastrophe may serve as a stark reminder of things to come. Yet, it is the prospect of such a civil war that has also removed the necessity for open Russian military intervention. Russia has begun to fight the war by proxy, by using covert military operatives and/or mercenaries.[44] Reflecting these developments and having nothing further to gain from an invasion, Russia announced the temporary withdrawal of regular combat troops from the border in June.

After adopting a 'retro' USSR foreign policy,[45] Putin needed and found new strategic allies: in May 2014, he entered into a gas deal with China,[46] which has the potential not only to disrupt vital energy supply to Europe but also to question the emergence of a future long-term cooperation based on mutual economic interest and trust. Whether these developments herald the coming of a new 'cold war' remains to be seen. What is evident, though, is that the Cold War's 'Strategic Stability' dogma, which prevented any direct military confrontation between NATO and the Soviet-led Warsaw Pact, does not exist in the 21st century. New technologies in 'cyber-space'[47] and the use of 'new wars' along asymmetric lines of conflict—'hybrid war' will see to it. Russia's operation has also shown that the hybrid approach can be adopted by states as well and not only by NSAs in an asymmetric context. In fact, it seems as if a resourceful state can wage hybrid war very effectively against opponents who lack the same resources. For example, one can look at the media advantage, which Russia had against Ukraine, a media advantage that is very much the backbone of the Russian new way of waging war. Once again, we have to remind ourselves that this media component is not a mere side-effect any more but the very core of post-industrial warfare.

The failure to agree on effective and far-reaching economic sanctions against Russia has also highlighted the weakness of the globalized economical system as such. But does the Crimean scenario teach something new about warfare? Some researchers have focused on the conventional part of the operation.[48] But it seems that the use of a term like 'semi-covert operations' in such texts is just a placeholder for a more accurate term such as 'hybrid war.'[49] Others have focused on what is new in Russian warfare, something about which Russia is very explicit. Among a host of features of the new war, there are some worrying elements we would like to consider: the non-declaration of war, the use of armed civilians, non-contact clashes like the blockade of military installations by 'protestors,' the use of asymmetric and indirect methods, simultaneous battle on land, air, sea, and in the informational space, and the management of troops in a unified informational sphere.[50]

Why bother with all these methods, as Russia can be strong enough to take on whatever Russia is interested in within its sphere of interest? Seen from the perspective of hybrid warfare, it is all about muddling Clausewitz's dictum of war as the continuation of politics with other means: no war was to be declared officially and civilians were to be used instead of combatants. What we have seen in the Crimea is that Russia acted very much in this way, actually denying the existing of a state of war but defining military action in a holistic way with armed as well as unarmed civilians, supported by regular combat elements, doing the actual military maneuver acting. The nature of the conflict remains undefined to a certain extent: war or civil unrest, interstate aggression or intrastate conflict. The latter was especially true in eastern Ukraine where the situation was very unclear when it came to whether Russia actually was active or not in an instrumental way. Against that backdrop, the following has become reality:

> With the advent of hybrid threats we will redefine what war is and we will most likely go into an era when we must get used to war and all its implications on society, there will possibleb [sic] be no difference between mission area and at home anymore, nor will the boundary between war and peace be well defined. 'Normality' will thus be redefined accordingly in a radical way.[51]

The international community and *jus ad bellum* are oriented towards limiting the possibilities of action in regular conflicts as we have come to know them in the 20th century. The hybrid logic of practice effectively amends the rules of war. Further, the practice of not acknowledging one's own actions makes the legal liability a difficult issue.

Africa

In Ukraine and the Crimea, we have seen Russia utilizing the hybrid approach. This is a bit of a novelty as when the term emerged at first it was a way of describing a non-state approach, namely Hizbollah in Lebanon in 2006. One could argue whether the term 'hybrid threats' can still be applied on NSAs, if one lays claims, that what we have seen in the Ukraine, is a hybrid conflict between Ukraine and the Russian separatists. On the other hand, one has to look at the logic of practice in every conflict in order to determine what the indicators are. It is of course important to note whether an actor is a state or not.

But which kind of indicators do we find in Boko Haram and Al-Shabaab that we can see as rather new and within the discussion of hybrid threats? For radical Islamists, the religious and political representations of the West do not match the attitude of their own culture towards a non-material rationality breaking through in the West with the Enlightenment. The Western world and its secularization serve more like a warning example for these groups. In any case, the rise of the radical Muslim movements can be seen as a reaction to modernism. Is an upsurge of Islam a form of neo-conservatism? This is an empirical question which will have to wait for now. But many of the insurgents in Boko Haram and Al-Shabaab come from countries where there is little room for anything else than radicalization when it comes to political room within which to maneuver.

Something that should be taken into consideration is that it is rather prejudiced to view all forms of religion as a quest for the past. It is possible and often the case, that religion defends the past. But it is also possible to imagine a progressive religious movement that, much like postmodernism, embraces and builds on—rather than repels—the movement that it reacts to, a concept which will be further explored later on when presenting examples of contemporary Islamist movements. Either way, both Islamism and postmodernism can be seen as reactions to a modernism that culminates in a globalisation and weakened national states. The trigger of this culmination was the end of the Cold War. Religion can provide existential comfort in an ever-changing world in a more striking way than postmodernism.

Which similarities between the events in the Crimea and the African theatres of terrorism can we then identify? Are there any similarities of Russia's conduct of operations and NSAs such as Boko Haram and Al-Shabaab? The most important similarity is the urge for media recognition, as proper media attention is crucial in the age of modern mass media communication. Is there something in common between the Crimean and Kenyan/Nigerian scenarios? Is it the same, and are both hybrid wars? In our perspective, 'hybrid threats' is a term which should be the litmus test of what future conflicts are to present to us as our immediate future reality. Yes, that is true, both scenarios use media as an integral part of warfare, not just as a collateral effect of the belligerent actions. But in the African radical Islamist cases, we see the same pattern as we have seen in, for example, Iraq both now under ISIS and under the insurgency against the USA. The difference in the use of media is fundamental. Radical Islamists use media as a 'force multiplier' for their terrorist agenda. In the Crimean case, we saw a plethora of misinformation: from the tactical level up to the Russian president, trying their best to communicate strategically that they were not involved in the operations while actually being caught red-handed. Even when three tanks crossed the border from Russia to Ukraine, none of the actors stated that they were the perpetrators and Ukraine did not push the point against Russia either.[52] The common denominator is the use of media in a very central role. The difference is that Boko Haram and Al-Shabaab try to translate tactical success into terror, while in the Crimean and Ukrainian examples, Russia tried the opposite while denying being an active agent. In the former case, *jus ad bellum* is ignored; in the latter, it is evaded.

This article was written with the intention of making 'hybrid threats' as a 21st-century security threat known to the wider audience despite NATO's decision not to adopt a comprehensive approach. This failure does not reduce the dangers of this category of global risks. Ongoing debate and academic engagement with the topic and rationale of 'hybrid threats,' such as the Swedish experiment in 2012, will hopefully lead to further awareness and eventually preparedness. This submission concludes with a sobering prediction: it is the opinion of the authors that the present legal concepts on the use of military force, the *jus ad bellum*, have become relatively anachronistic and even partially outdated, something that will not suffice when dealing with the security threats and challenges of the 21st century. The authors predict that the emergence of hybrid threats and their recognition as potential threats to peace and security as such, the proliferation of low-threshold regional conflicts (such as the 2011 Libyan conflict, Syria and now Iraq), as well as continuing asymmetric warfare scenarios (such as Syria, Iraq, Afghanistan, and Pakistan) will have a significant influence on the prevailing culture and prism of traditional military activity, which is still influenced by concepts from the previous century. With such a change of military doctrines, a change of legal paradigms will be inevitable: new adaptive means and methods of 'flexible responsiveness' through escalating levels of confrontation and deterrence will question the existing legal concept of the prohibition of the use of force with its limited exceptions, as envisaged under Articles 2(4) and 51 of the UN Charter and Article 5 of the NATO Treaty.[53] Future direct intervention in failed state scenarios will require flexibility in terms of choice of military assets and objectives. Future responses to multi-modal threats will always include the kinetic force option, directed against—most likely—NSAs. They will also affect our present concepts of the illegality of the use of force in international relations, as enshrined in Article 2(4) of the UN Charter with limited exceptions available under Article 51 of the UN Charter, namely individual and collective self-defense (*cf.* Article 5 of the NATO Treaty) as well as UN authorization. Already today, the continuing use of UAVs (unmanned aerial vehicles, or drones) for 'targeted killing' operations effectively emphasizes the legal challenges ahead: the ongoing 'kill' operations in the so-called 'tribal' areas of Waziristan/Pakistan are kinetic military operations, which demonstrate how quickly the critical threshold of an armed conflict can be reached and even surpassed. These operations clearly fall within the scope of the definition of 'armed conflict' by the International Criminal Tribunal for the former Yugoslavia in the appeal decision in *The Prosecutor v Dusko Tadic*[54] and therefore giving rise to the applicability of the norms of the so-called 'Law of Armed Conflict,' the body of international humanitarian law governing conduct in war. The 'lawfulness' of such operations does, however, require the existence of either a mandate in terms of Article 51 of the UN Charter (in the form of a United Nations Security Council [UNSC] Resolution authorizing the use of force in an enforcement and peace enforcement operation context) or the existence of an illegal armed attack in order to exercise a right to national or state self-defense in terms of Article 51 of the UN Charter. Whether such military operations are within the scope of these categories remains open to discussion.

NATO's Strategic Concept of 2010 was aimed at prevention as well as deterrence in general and at developing a holistic or comprehensive approach to a variety of new conflict scenarios of multi-modal or hybrid threats, from kinetic combat operations to multi-stakeholder-based non-kinetic responses. Even with the failure to formulate a binding comprehensive approach to such threats at the supranational level, the findings of NATO's hybrid workshops have shown the significance of such threats and the need to respond in a flexible way.

New roles of states, their militaries and their politicians but also NSAs, such as multinational corporations and non-governmental organization (NGOs), are needed. Geography as a term has already become obsolete as the 'war on terrorism' has shown: with its abstract categories of distinction into 'abroad,' such as 'mission area,' 'area of operations' and 'theatre of operation,' and 'at home' having merged into one abstract universal 'battlefield' with an often-shifting geographical dimension. The dogma of 'flexible response,' which has often been regarded as a tenet in military operational thinking and doctrine, has lost much of its meaning as a means of military force projection within the context of hybrid threats.

Hybrid threats pose not only security challenges but also legal ones and only time will tell how Western societies with their military will eventually adapt within their existing legal and operational frameworks.

Endnotes

1. The term 'cyber' is used in a wider sense, referring to the use of computer technology and the internet for operations in the so-called fifth dimension; 'cyber operations,' 'cyber war,' and 'cyber attacks' are examples of such operations, depending on their intensity. For a classification of 'cyber conflicts,' see Schmitt, M. "Classification of cyber conflict." *Journal of Conflict & Security Law* 17/2. 2012. 245–260.
2. Lamp, N. "Conceptions of war and paradigms of compliance: The 'new war' challenge to international humanitarian law." *Journal of Conflict & Security Law* 16/2. 2011. 223.
3. Hoffman, FG. "Hybrid threats: Reconceptualizing the evolving character of modern conflict." *Strategic Forum* 240. 2009. 1;

also see Hoffman, FG. "Hybrid warfare and challenges." *Joint Forces Quarterly* 52. 1Q. 2009. 1–2; Hoffman, FG. "Hybrid vs. compound war: The Janus choice of modern war: Defining today's multifaceted conflict." *Armed Forces Journal* October 2009. 1–2.

4. The authors have undertaken some prior work in that field, which reflects on various other aspects of the topic; *cf* Bachmann, S-D & Kemp, G. "Aggression as 'organized hypocrisy:' How the war on terrorism and hybrid threats challenge the Nuremberg legacy." *Windsor Yearbook of Access to Justice* 30/1. 2012; Bachmann, S-D. "NATO's comprehensive approach to counter 21st century threats: Mapping the new frontier of global risk and crisis management." *Amicus Curiae* 88. 2011. 24–26; Bachmann, S-D & Gunneriusson, H. "Countering terrorism, asymmetric and hybrid threats: Defining comprehensive approach for 21st century threats to global risk and security." Swedish MoD—High Command, Internal Paper, releasable to the public some notions of this article were published prior in another context in "Terrorism and cyber attacks as hybrid threats: Defining a comprehensive approach for countering 21st century threats to global peace and security." *Journal for Terrorism and Security Analysis* 2014. 26–37.

5. Hoffman, FG. *Conflict in the 21st century: The rise of hybrid wars.* Arlington, VA: Potomac Institute for Policy Studies, 2007, 37.

6. See for example Matthews, MM. "We were caught unprepared: The 2006 Hezbollah-Israeli War." The Long War Series Occasional Paper No 26. Fort Leavenworth, KS: US Army Combined Arms Center, Combat Studies Institute Press, 2008.

7. (n 4) *supra*.

8. *cf* BI-SC input for a new NATO Capstone Concept for the Military contribution to countering hybrid enclosure 1 to 1500/

CPPCAM/FCR/10-270038 and 5000 FXX/0100/TT-0651/SER: NU0040, 25 August 2010.

9. Miklaucic, M. "NATO countering the hybrid threat." 23 September 2011. <http://www.act.nato.int/nato-countering-the-hybrid-threat> Accessed on 7 May 2015.

10. For a thorough discussion of the concept of hybrid threats, see Sanden, J & Bachmann, S-D. "Countering hybrid eco-threats to global security under international law: The need for a comprehensive legal approach." *Liverpool Law Review* 33. 2013. 16. Copyright of the authors is acknowledged; Aaronson, M, Diessen, S & De Kermabon, Y. "Nato countering the hybrid threat." PRISM 2/4. 2011. 115.

11. A programme of practical bilateral cooperation between individual Euro-Atlantic partner countries and NATO.

12. Försvarshögskolan. "Hur försvarar vi oss mot hybridhot?" 30 October 2012. <https://www.fhs.se/sv/nyheter/2012/hur-forsvarar-vi-oss-mot-hybridhot> Accessed on 27 January 2014.

13. A recent example highlights the problem of failed coordination between the Secret Police, the National Defence Communication and Military Intelligence. The official in charge did leave office, as it was little to coordinate; Sverigesradio. "Spionchefer lämnar samarbetsorgan i protest." 4 November 2012. <http://sverigesradio.se/sida/artikel.aspx?programid=83&artikel=5334446> Accessed on 27 January 2014.

14. Feedback from scenario participants (answers to Gunneriusson's call for feedback by email): University representative 1; Medicine sphere representative 1; Armed Forces representative 3; Cyber security representative 1 and the Swedish Defence Materiel Administration representative 2.

15. *Ibid.*

16. Olearchyk, R & Buckley N. "Ukraine's security chief accuses Russia of waging 'hybrid war.'" *The Financial Times*. <http://www.ft.com/cms/s/0/789b7110-e67b-11e3-9a20-00144feabdc0.html#axzz33DXeBUkR> Accessed on 6 May 2015.

17. Bachmann, S-D. "Crimea and Ukraine 2014: A brief reflection on Russia's 'protective interventionism.'" *Jurist*. <http://jurist.org/forum/2014/05/sascha-bachmann-ukraine-hybrid-threats.php> Accessed on 6 May 2015.

18. SNDC Hybrid Threat Workshop, Swedish Armed Forces representative.

19. On biohacking, see Ricks, D. "Dawn of the BioHackers." *Discover*. <http://discovermagazine.com/2011/oct/21-dawn-of-the-biohackers/article_view?b_start:int=2&-C=> Accessed on 27 January 2014 and Saenz, A. "Do it yourself biohacking." *SingularityHUB*. 28 April 2009. <http://singularityhub.com/2009/04/28/do-it-yourself-biohacking/> Accessed on 27 January 2014; Whalen, J. "In attics and closets, 'biohackers' discover their inner Frankenstein." *Wall Street Journal*. 9 May 2012. <http://online.wsj.com/article/SB124207326903607931.html> Accessed on 27 January 2014.

20. *dn.se*. "Säpo: Högskoleutbildningar kan sprida kärnvapen." 10 April 2014. <http://www.dn.se/nyheter/sverige/sapo-hogskoleutbildningar-kan-sprida- karnvapen/> Accessed on 7 May 2015.

21. See in general Döge, J. "Cyber warfare: Challenges for the applicability of the traditional laws of war regime." *Archiv des Völkerrechts* 48. 2010. 486.

22. See Traynor, I. "Russian accused of unleashing cyberwar to disable Estonia." *The Guardian*. 17 May 2007. <http://www.guardian.co.uk/world/2007/may/17/topstories3.russia> Accessed on 12 May 2015.

23. Marcus, L. "Explosive new Arab music video: 'Strike a blow at Tel Aviv.'" <http://www.jewishpress.com/special-features/

israel-at-war-operation-amud-anan/explosive-new-arab-music-video-strike-a-blow-at-tel-aviv/2012/11/19/> Accessed on 15 May 2013.

24. Jaber, H. "Hamas goes underground to avoid drones." *The Sunday Times.* 25 November 2012. 27.

25. Halliday, J. "WikiLeaks: US advised to sabotage Iran nuclear sites by German thinktank." *The Guardian.* 18 January 2011. <http://www.theguardian.com/world/2011/jan/18/wikileaks-us-embassy- cable-iran-nuclear> Accessed on 6 May 2015; Kroft S. "Stuxnet: Computer worm opens new era of warfare." *60 minutes.* 4 June 2012. <http://www.cbsnews.com/news/stuxnet-computer-worm-opens-new-era-of-warfare-04-06-2012/> Accessed on 6 May 2015; Williams, C. "Stuxnet: Cyber attack on Iran 'was carried out by Western powers and Israel.'" *The Telegraph.* 21 January 2011. <http://www.telegraph.co.uk/technology/8274009/Stuxnet-Cyber-attack-on-Iran-was-carried-out-by-Western-powers-and-Israel.html> Accessed on 6 May 2015.

26. With the decision taken in 2009, and initial operational capability as of 2010, see US Strategic Command. "US Cyber Command." <http://www.stratcom.mil/factsheets/2/Cyber_Command/> Accessed on 6 May 2015.

27. *Ibid.*

28. See NATO. "Transformation Network." <https://transnet.act.nato.int/WISE/TransformaI/ACTIPT/JOUIPT> Accessed on 12 May 2015.

29. Some of the following content derives from Swedish National Defence College sources, which are on file with the authors.

30. Jagoindia. "Chilling phone transcripts of Mumbai terrorists with their Lashkar handlers." 7 January 2009. <http://islamicterrorism.wordpress.com/2009/01/07/chilling-phone-transcripts-of-mumbai-terrorists-with-their-lashkar-handlers/> Accessed on 27 January 2014.

31. Paganini, P. "Malware hides C&C server communications using Google Docs function." 21 November 2012. <http://securityaffairs.co/wordpress/10454/malware/malware-hides-cc-server-communications-using-google-docs-function.html> Accessed on 27 January 2014.
32. Baloch, F. "Cyber vandalism: Hackers deface Google Pakistan." *The Express Tribune*. 25 November 2012. <http://tribune.com.pk/story/470924/cyber-vandalism-hackers-deface-google-pakistan/> Accessed on 27 January 2014.
33. *Proofpoint*. "Proofpoint Uncovers Internet of Things (IoT) Cyberattack." 16 January 2014. <https://www.proofpoint.com/us/proofpoint-uncovers-internet-things-iot-cyberattack> Accessed on 20 January 2014.
34. Jānis. B. "Russia's new generation warfare in Ukraine: Implications for Latvian defense policy." Policy Paper No 02. Riga: National Defense Academy of Latvia, 2014, 9.
35. See for example The White House. "The National Security Strategy of the United States of America" (hereafter NSS). September 2002. <http://nssarchive.us/NSSR/2002.pdf> Accessed on 15 May 2013, reaffirmed in NSS 2012.
36. . . . and often questionable in terms of legality and legitimacy, and might qualify as the prohibited use of force in terms of Article 2(4) of the UN Charter. The planning and conducting of these operations would in the future fall within the scope of Article 8 *bis* of the ICC Statute (in its revised post-Kampala 2011 version and coming into force only after 2017, potentially leading to individual criminal responsibility).
37. The ideas of the extreme Wahhabism (the religious fundament advocated by al-Qaeda), that man should live in the same technological conditions as Muhammad, is easily linked to the ideas behind fourth-generation warfare.

38. UNODC. "The use of the internet for terrorist purposes." <http://www.unodc.org/documents/frontpage/Use_of_Internet_for_Terrorist_Purposes.pdf> 13 Accessed on 27 January 2014.
39. Clausewitz, C. *On war* (Book I), transl Graham, JJ. London: N Trübner, 1873, 24; Braudel, F. *La Méditerranée et le monde méditerranéen à l'époque de Philippe* (Vol. II). Paris: Lib. A. Colin, 1949.
40. *Ibid.*
41. Sun-Tzu. *The art of war* (transl L Giles). New York: Barnes & Noble, 2012, Chapter 4.
42. See Bachmann, S-D. "Russia's 'spring' of 2014." *OUPblog.* 9 June 2014. <http://blog.oup.com/2014/06/russia-putin-hybrid-war-nato/> Accessed on 6 May 2015 (copyright of OUP fully acknowledged).
43. Varenytsia, I. "Fighting in Eastern Ukraine rages on overnight despite talks." *Time.* 14 April 2015. <http://time.com/3820764/ukraine-donetsk-russia-fighting-rebels-belarus-putin-shyrokyne/> Accessed on 7 May 2015.
44. Roth, A & Tavernise, S. "Russian revealed among Ukraine fighters." *The New York Times.* 27 May 2014. <http://www.nytimes.com/2014/05/28/world/europe/ukraine.html?_r=0> Accessed on 7 May 2015.
45. Patel, A. "Russia with Crimea: Back in the USSR." *Rusi.* 9 May 2014. <https://www.rusi.org/analysis/commentary/ref:C536CCA853F88D/#.U4irg kp-58E> Accessed on 7 May 2015.
46. Lain, S. "Russia's gas deal with China underlines the risks to Europe's energy security." *The Guardian.* 26 May 2014. <http://www.theguardian.com/commentisfree/2014/may/26/russia-gas-deal- china-europe-energy-security-danger> Accessed on 7 May 2015.

47. Roscini M. "Is there a "cyber war" between Ukraine and Russia?" <http://blog.oup.com/2014/03/is-there-a-cyber-war-between-ukraine-and- russia-pil/> Accessed on 6 May 2015.

48. Sutyagin, I & Clarke, M. "Ukraine military dispositions: The military ticks up while the clock ticks down." Briefing Paper. *RUSI*. April 2014. <http://www.rusi.org/downloads/assets/UKRANIANMILITARYDISPOSITIONS_RUSIBRIEFING.pdf> Accessed on 6 May 2014; Norberg, J. "The use of Russia's military in the Crimean crisis." *The Global Think Tank*. 14 March 2013; Norberg J. "The use of Russia's military in the Crimean Crisis." *Carnegie Endowment for International Peace.* 13 March 2014. <http://carnegieendowment.org/2014/03/13/use-of-russia-s-military-in- crimean-crisis/h3k5?reloadFlag=1> Accessed on 6 May 2015.

49. Norberg, J & Westerlund, F. "Russia and Ukraine: Military-strategic options, and possible risks, for Moscow." *IISS*. 7 April 2014. <https://www.iiss.org/en/militarybalanceblog/blog sections/2014-3bea/april-7347/russia-and-ukraine-3b92> Accessed on 6 May 2015.

50. Changes in the Character of Armed Conflict According to General Valery Gerasimov, Chief of the Russian General Staff, listed in Jānis *op. cit.*, p. 4

51. Gunneriusson, H. "Nothing is taken serious until it gets serious." *Defence Against Terrorism Review* IV/7. 2012.

52. Moore, J. "Ukraine crisis: Three Russian tanks cross shared border." *International Business Times*. <http://www.ibtimes.co.uk/ukraine-crisis-three-russian-tanks-cross-shared-border-1452424> Accessed on 6 May 2015.

53. See Bachmann & Kemp *op. cit.*, n 4, for a detailed overview of possible legal challenges in the context of kinetic responses to hybrid threats.

54. IT-94-1-A, 105 *ILR* 419,488.

Sascha-Dominik Bachmann
Assessor Jur, LLM (Stel) LLD (UJ), Associate Professor in International Law (Bournemouth University, UK). Outside academics, he has served in various capacities as lieutenant colonel (army reserve), taking part in peacekeeping missions in operational and advisory capacities. The author took part as NATO's Rule of Law Subject Matter Expert (SME) in NATO's Hybrid Threat Experiment of 2011 and in related workshops at NATO and national level.

Håkan Gunneriusson PhD
Modern History, Associate Professor in War Studies, head of research ground operative and tactical areas Department of Military Studies, War Studies Division, Land Operations Section, Swedish Defence University.

4

Is Cyber Shape-Shifting?

Neal Kushwaha; Founder and CEO, Impendo Inc.
Bruce W. Watson; Chief Scientist, IP Blox

Abstract

Technologies have evolved so rapidly that companies and governments seem to be regularly trying to catch up to new capabilities and thereby making quick decisions that have the potential to set precedents and present international challenges.[1] Is cyber capability changing so fast that our sensemaking is lagging? Is cyber shape-shifting?

With the opportunity to take a step away from the technical aspects of cyber and consider a cyber taxonomy, this paper explores the domain of cyber by structuring the conceptual problems and by putting the individual small solutions into their respective places within a conceptual framework.

The paper breaks cyber into seven (7) concepts and discusses each of them:

1. knowledge trajectory – aligning cyber to knowledge economies;
2. discrimination – categorizing various cyber weapons;

1 The Joint Statement for the Record to the Senate Armed Services Committee, Foreign Cyber Threats to the United States (January 5, 2017), p5 paragraph 1 states "*...countries do not widely agree on how such principles of international law as proportionality of response or even the application of sovereignty apply in cyberspace.*" (Clapper, Lettre, & Rogers, 2017)

3. recombinant and mutable – discussing how cyber weapons can be easily modified when compared to traditional kinetic weapons;
4. model/object dichotomy collapse and free replication – discussing how in cyber, the code is the object, making it easy to duplicate the weapon and how traditional methods of sanctions may no longer be suitable;
5. speed of light – the challenge of detecting cyber weapons and the ease with which they can be shared;
6. dynamic multidimensional space – discussing the change in theater of operations and how collateral damage is an expected outcome; and
7. scope of impact – discussing the true impact of cyber weapons and their behavior.

This position paper challenges the reader further with the radical possibility that cyber is not a Domain of Warfare and that the term "cyber attack" may likely benefit from an alternate label such as "cyber espionage" or "cyber sabotage." We discuss how cyber is impaired by:

1. attribution, making it difficult to identify the source;
2. scope of impact resulting in manipulation, interruption/disruption, and bullying; and
3. high dependence on the target's cyber hygiene and IT business processes.

Because of these challenges, we propose that cyber is for now rather a tool or tradecraft for the purpose of espionage or sabotage.

Introduction

In the Cyber Domain, nearly every criminal act is described as an "attack." Over the past few decades, numerous claimed "cyber attacks" have been carried out by various sources with various degrees of impact and using various vectors of attack.[2] But is it suitable and widely accepted to classify these attacks[3] in cyber as in a Domain of Warfare?[4]

The evolution of technology on both the hardware and software sides has been widely embraced around the globe. The prevalence of Internet-based services and the desire to be constantly connected to one another is an unstoppable energy. Although a war could take place in this space, the decision to declare this space as a Domain of Warfare must not be taken hastily. One may be able to injure or kill one another with a fork or a spoon, but it most likely does not mean we create a Domain of Warfare for Utensils.

NATO[5] (defensive only) and various nations[6] have already stood up Cyber Operations[7] and/or Cyber Commands[8] in light of the trend,

---

2 Middleton (Middleton, 2017), NATO Review Magazine (NATO Review Magazine, 2013), and Vaidya (Vaidya, 2015), each describe the history of cyber crime activities.

3 In Schmitt's paper, he examines the meaning of "*attack*" (Schmitt, 2012) as it applies to Cyber Operations and International Law.

4 On pp680-681 in chapter "*Cyber Warfare*" of Solis' textbook, he discusses the definition of "*cyber attack*" and related behaviors under the Law of Armed Conflict (LOAC) and International Humanitarian Law (IHL). It states: "*For both international and noninternational armed conflict, an excellent definition of a cyber attack is a cyber operation, whether offensive or defensive, that is reasonably expected to cause injury or death to persons, or damage or destruction to objects.*" (Solis, 2016, pp. 680-681)
The subsequent paragraph states: "*...cyber theft, cyber intelligence gathering, and cyber intrusions that involve brief or periodic interruption of nonessential cyber services clearly do not qualify as cyber attacks.*" (Solis, 2016, pp. 680-681)

5 Paragraphs 5 and 6 describes "*NATO's [cyber] defensive mandate*" and that the allies "*recognise cyberspace as a domain of operations in which NATO must defend itself as effectively as it does in the air, on land, and at sea*" (Minárik, 2016).

6 Joint Statement for the Record to the Senate Armed Services Committee (January 5, 2017), p5 paragraph 2, "*As of late 2016, more than 30 nations are developing offensive cyber attack capabilities.*" (Clapper, Lettre, & Rogers, 2017)

7 Communications Security Establishment Act, subsection Mandate, paragraph 19 "*Defensive cyber operations*" and paragraph 20 "*Active cyber operations*" (House of Commons of Canada, 2017)

8 In paragraph 1 of a memo to the Secretary of Defense, President Trump states: "*...I direct that U.S.*

considering cyber or cyberspace as a Domain of Warfare among land, maritime/sea (including surface and subsurface), air, and space. Considering that cyber spans all domains, the driving governance behind these organizations or branches must be well defined.

In an effort to give cyber a proper comparison to traditional kinetic warfare, this paper explores a taxonomy of cyber to help provoke thought. Although the paper does not deliver a definitive conclusion or present qualitative test results, it does encourage the reader to study the presented concepts in a different way that may change the way we perceive and consider cyber operations. To tackle this task, the paper discusses seven (7) concepts that encompass the cyber domain. The concepts labeled in the mind map of Figure 1 below are the header for each section of this paper. Each concept of cyber branches further into a hierarchy of ideas that are discussed in detail throughout this paper.

Figure 1. Mind Map of Cyber Operations

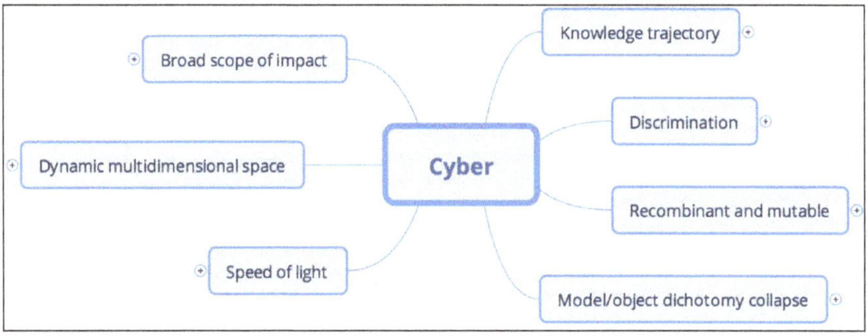

Knowledge Trajectory

The maturity of any discipline can be described as a trajectory of knowledge. When exploring the cyber domain, we can appreciate that only a few, if any, have mastered it. Table 1 below shows the five stages of knowledge economies.[9]

Cyber Command be established as a Unified Combatant Command." (Trump, 2017)

9 Knowledge economies are described by Dr Watson (Watson) and that work in itself is an extension of the discussion in the book "*Software Architecture: Perspectives on an Emerging Discipline*" (Shaw & Garlan, 1996).

Art: Each artist has their own representation of the scene, resulting in significantly different paintings. Advanced persistent threats (APTs) are a form of an art. It takes a skilled resource that has rare and unique talents to be able to create APTs. Each APT behaves differently from another when applying their unique talents.

Craft: Groups such as those with Computer Incident Response Capability (CIRCs)[10] and those with Computer Emergency Response Teams (CERTs)[11] and those that create cyber weapons tend to depend on a model of sharing information. The notion of master and apprentice is normally applied to sharing a craft and can be seen in cyber when a student learns from a teacher using online videos, thereby creating consistent teaching. These groups of masters and apprentices create towers of knowledge and/or styles.

Discipline: Cyber defence systems and the practice of cyber hardening rely on heuristics or "rules of thumb." Hackers can also fall into this group whereby they download tools and follow a set of common practices to achieve their goal. Users of these tools may not necessarily understand the underlying defensive cyber operations methods or active/combative cyber operations exploits, but they are able to use the tools in keeping with the rules of thumb.

Science: Science encompasses the deep theoretical foundations of the rules of thumb present in a discipline. When it comes to cyber, this falls under Computer Sciences. Individuals or groups of individuals who write malicious code or derive methods to perform cyber crimes are examples of the use of science in cyber. Reversing or dissecting of code, along with research and testing to cultivate vulnerabilities and exploits, are other applications of science.

Engineering: Finally, the knowledge of engineering advances common practices into best practices. The result of successful engineering in cyber is the industrialization of a cyber weapon.

---

10 Examples of CIRCs: Canada Cyber Incident Response Centre (Public Safety Canada, 2016) and NATO Computer Incident Response Capability (NATO Communications and Information Agency, 2016)

11 Examples of CERTs: US (US-CERT), Carnegie Mellon University (Carnegie Mellon University CERT®), and Australia (Australia Computer Emergency Response Team (CERT)

When these five Knowledge Economies are aligned with the Capability and Maturity Model Integration (CMMI Institute) from 1 to 5 (respectively, chaotic, repeatable, defined process, managed, optimization, and continual improvement), we can assign the cyber maturity of various individuals, companies, and governments in an unconventional qualitative manner.

Table 1. Knowledge Economies Related to Cyber and CMMI

Knowledge Economies	Definition	Relation to Cyber	CMMI
Art	• Rare unquantified talent • Wide range of consistency	• Advanced persistent threats	Initial or chaotic
Craft	• Master and apprentice • Consistency from good teaching	• CIRC and CERT • Cyber weapon construction (or the writing of malicious code)	Managed or repeatable
Discipline	• Heuristics	• Hacking and Cyber defence • Practice of cyber hardening	Defined
Science	• Underlying principles are well understood	• Cyber as part of computer science	Qualitatively managed
Engineering	• Best practices applied • Highly reproducible	• Kinetic warfare	Optimizing or continual improvement

Since we choose to be connected over the Internet and suffer from each other's shortcomings and failures, we may collectively be very low on the maturity scale even if we individually measure higher using the knowledge economy model.

Discrimination

In general, cyber weapons can be organized into four high level categories: weapons of mass interruption, weapons of mass manipulation, surgical or tailored weapons, and weapons of mass destruction. Each of these four categories is described below.

A. *Weapons of Mass Interruption (WMI)*

Weapons of mass interruption cause a form of chaos in the

interconnected cyber space by introducing delays into normal cyber workflow. Well-known examples of weapons of mass interruption are, but not limited to, wide spreading viruses, Trojans, worms, botnets, or email spamming resulting in a variety of outcomes including diminished or denied services to exfiltration of data including passwords. The offending operations often include a relatively small introduction or alteration of information on a system resulting in an increased amount of network traffic from the infected system.

Some historical examples of this type of weapon include the following below.

1. 2000-05-04: ILOVEYOU (worm) (CERT®, 2000)
2. 2001-09-18: Nimda (worm) (CERT®, 2001)
3. 2003-01-24: SQL Slammer (worm) (CERT®, 2003)
4. 2004-01-26: Mydoom (virus) (CERT®, 2004)
5. 2007-01-19: Storm (Trojan) (Symantec, 2007)
6. 2009-03-29: Confiker (worm) (US-CERT, 2013)
7. 2009-05-28: Bredolab (Trojan) (Symantec, 2012)

Interestingly, the impact of weapons of mass interruption have not been as prevalent in the recent past.[12] One may say it could be credited to our improved level of maturity that enables us to detect and defend against such attacks with common tools. One may also argue that it may be a result of countermeasures and that countermeasures trigger more opportunities to counter from all sides; however, this behavior is unlikely as we still see some of these types of weapons reaching cyber borders.[13]

---

12 Wikipedia, article on "*Timeline of computer viruses and worms*" (Wikipedia, 2017), when reviewing the list of malware in the timeline presented, one can see the progression from WMI to WMM. The article lists WMI malware in 2015 through 2017; however, the quantity is far less than those described in the 1990 through 1999 and 2000 through 2009 inclusive.

13 Benzmüller's review of AV-Test's statistical data of new malware between 2007-2017 revealed that in 2016 there were on average *"780 [new malware specimen] per hour"* and in 2017Q1 alone there were on average *"858 [new malware specimen] per hour."* (Benzmüller, 2017)

B. Weapons of Mass Manipulation (WMM)

Weapons of mass manipulation tend to alter or delete information. "Cyber attacks" using weapons of mass manipulation generally have longer-lasting effects than those causing interruption; however, a well-managed and mature Information Technology service delivery model can generally recover within a short and/or reasonable amount of time, and with this capability, the weapon behaves akin to a weapon of mass-interruption technique versus mass manipulation.

Recovery from mass manipulation may involve the complete destruction or quarantine of the attacked systems followed by a full restoration of data using a safe and uninfected backup set.[14] Hardware impacted by firmware-altering attacks can prove to be more time-consuming to recover from, as the ability to restore firmware may not always be possible, thereby resulting in potential hardware procurement and shipment timeline challenges.

Ransomware is a good example of a manipulation weapon. It alters the data on a computer using a method of encryption but does not delete it immediately. Although the prevalence of ransomware seems to be rising as of mid-2016 through 2017, in comparison to other types of malware and cyber weapons, it remains a negligible quantity.[15]

An example of a weapon of mass manipulation was the Saudi Aramco deletion of data and overwriting of the Mast Boot Record (MBR) of over 35,000 computers, via the Shamoon virus. In 2012, Shamoon also left behind propaganda showing what was likely supposed to be a burning flag of the United States of America (USA); however, possibly due to poor coding, the burning flag was only partially visible (Wikipedia, 2017).

---

14 Symantec's "*Ransomware: 5 dos and don'ts*" describe (1) the proper cyber hygiene including performing backups and (2) the behavior of common ransomware. (Symantec, 2017)

15 Benzmüller summarized the ransomware analysis of 2016 through 2017 as "*...the total volume of ransomware was hardly detectable and vanishes in the flood of other malware.*" He also noted that "*The share of ransomware is growing substantially. In the general flood of malware it is hardly measurable.*" (Benzmüller, 2017)

A second example of mass manipulation was the shutdown of thirty power substations in Ukraine also impacting small parts of other surrounding nations. The impact was short-lived, lasting between one and six hours. The hackers used a poorly-situated remote access point to the Supervisory Control and Data Acquisition (SCADA) network that bypassed air-gapped systems and the expected two-factor authentication.[16]

Weapons of mass manipulation are not only those that are inadvertently contracted; they can also be purposefully directed to result in the manipulation of public or political perception or opinion. These types of "cyber attacks" have commonly surfaced as fabricated articles, opinionated articles, and comments in social media circles to name a few.[17]

C. *Surgical or Tailored Weapons*

Surgical or tailored weapons are not discovered as easily, but when they are uncovered, they do receive a level of government and corporate attention.[18] They can be written and delivered to attack a specific unit of information technology at an organization level down to a specific file or object. Generally, they do not spread too far beyond the intended target and thereby limit collateral damage.

It is entirely possible that a tailored cyber weapon may be reuseable on another target without further customization of the weapon. This is likely due to the type of vulnerability or exploit that may have been used.

16 Zetter described the vector of attack in paragraph 8 of the article. (Zetter, 2016)

17 The ODNI describes some of the evidence accumulated by the CIA, FBI, and the NSA regarding Russia's (at a nation state level) campaign to influence the 2016 US Presidential election in the declassified report. [Office of the Director of National Intelligence (ODNI), 2017] Calabresi, paragraph 9 states "*...a Russian soldier based in Ukraine successfully infiltrated a U.S. social media group by pretending to be a 42-year-old American housewife and weighing in on political debates with specially tailored messages.*" (Calabresi, 2017) and "*...Russia created a fake Facebook account to spread stories on political issues like refugee resettlement to targeted reporters they believed were susceptible to influence.*" (Calabresi, 2017) Shane described the use of counterfeit online social media profiles, "*genuine accounts that had been hijacked*" (Shane, 2017), and the US national and international public that seem to be influenced by the Russian sourced individuals and stories.

18 Roberts discusses the use of SQL injection code to reveal unauthorized information of key organizations to hackers. (Roberts, 2017)

Examples of these are the theft of credit card details of SONY PlayStation and Microsoft Xbox clients (among others,)[19] the US Office of Personnel Management (OPM) data breach including personal information such as fingerprints of 5.6M federal employees (Wikipedia, 2017), Stuxnet on nuclear centrifuges in Iran (Mueller & Yadegari, 2012), and the BOTNET attack on Estonia (McGuinness, 2017), and the data breach at Equifax (Wikipedia, 2017).

D. Weapons of Mass Destruction (WMD)

At the time of writing, we have yet to be made aware of a cyber weapon of mass destruction, in other words, one that directly causes mass casualties and/or loss of life, damage to structures, or damage to the biosphere. One could present the case that the shutdown of thirty power substations affecting approximately 230,000 people during the 2015 "cyber attack" in Ukraine may have impacted lives; however, there is little reported evidence of this during the short-lived outage.

One may also consider the sabotage of the Siberian natural gas line explosion of 1982 (disputed as 1989) was the first documented "cyber attack." However, this has been contested as *"not caused by a system shutdown, but by deliberately creating overpressure in the pipeline by [manually] manipulating pressure-control valves in an active control process."* (Rid & McBurney, 2012, p. 9) Poor construction causing a leak followed by a poor decision to manually increase the line pressure may have caused the gas to ignite when two trains collided. (Wikipedia, 2017)

Transformers that feed hospitals can fail at any time, and the utility does not provide a redundant unit for them. We have seen rodents such as a squirrel take up shelter inside a transformer unit located on or near a facility causing a short resulting in a loss of utility power.[20] Replacing a failed primary transformer provided by

---

19 The Daily Mail article describes the acts of hackers known as "*LizardSquad,*" releasing 13,000 passwords and credit card details harvested from various companies on Christmas Day. (Boyle, 2014)

20 Mooallem, discusses various cases where squirrels are the root cause for the failure of power

the utility often has long lead times, sometimes as high as three to twelve months. During this time, the critical service such as a hospital is expected to operate with alternate power, such as uninterruptible power supplies supported by generators, flywheels, etc.

For data centers, the Uptime Institute defines utility as an "*economic alternative.*"[21] They also specify that data centers should expect loss of utility power and Tier III and IV facilities should live and operate through such conditions without any loss of critical load. Extending this to cyber, if a hospital or other critical service were to lose utility power due to a "cyber attack," it would be considered highly unlikely that they would not have an alternate source to generate their own power. If they did not have an alternate source, then the service could not be truly considered critical, as it was never prepared to support any type of loss of utility.

Alternatively, one may argue that the impact of a "cyber attack" to a state, organization, or even an individual can be a reflection of their cyber hygiene or negligence to remain current with global cyber security directions.

When it comes to air-gapped systems, the awareness to jump air-gapped networks and achieve access to command and control systems is likely not resting solely on cyber capability and involves the support of various other vectors of attack including but not limited to those such as HUMINT and SIGINT. This type of behavior is related to sabotage and espionage. (Wikipedia, 2017)

Recombinant and Mutable

Imagine the difficulty in conceptualizing and designing a traditional kinetic weapon that performs reasonably well in hand-to-hand combat and is equally capable of delivering a large

---

distribution. (Mooallem, 2013)

21 Uptime Institute under the section "*Utility feeds determine Tier level*" they state: "*...utility power is subject to unscheduled interruption—even in places with reliable power grids...Most Tier Certified data centers use utility power for main operations as an economic alternative, but this decision does not affect the owner's target Tier objective.*" (Uptime Institute, LLC, 2017)

explosive yield with a radius that could encompass an average European country. Now consider the same under cyber, where the programmer has merged existing code into a new cyber weapon that will be able to apply all actions of the code. The programmer does not need to design new components of the cyber weapon. They simply need to write a method to merge the capability under one package or payload of code.

The idea of hybridizing conventional kinetic weapons is something that is time consuming and challenging when attempting to achieve good results, however significantly simpler in cyber. An example of this is the merged SpyEye and ZueS malware. SpyEye impacted a variety of browsers while ZueS targeted Microsoft Windows environments. Both SpyEye and ZueS harvested banking user credentials. (Lyden, 2011)

From the mutable perspective, cyber weapons can also be written to change in battle, resulting in highly dynamic and agile weapons. Besides the ease of use, conventional weapons may be selected based on their yield among several other attributes; however, a cyber weapon can be equipped with access to a wide variety of dynamic alternatives that can be applied based on programming logic. These changes can be made very rapidly and designed to operate with or without human intervention.

Figure 2. Mind Map of Cyber Operations – Recombinant and Mutable

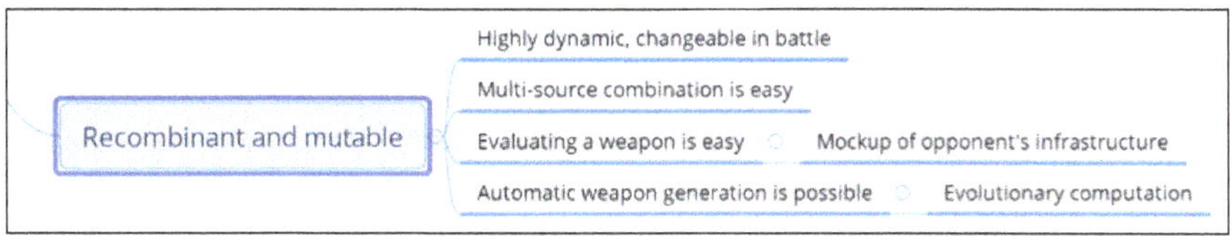

Evaluating cyber weapons can be discreet and also performed expeditiously. For example, testing a missile launch and detonation of payload can not only be time consuming but also very obvious to your adversaries and allies. In cyber, however, one could

create a simulated environment of an opponent's infrastructure and document the outcomes of the variants without informing other parties. Then, as needed, one could recreate the simulated environment in short order to perform more tests of the cyber weapon. Automated restoration of these logical virtual test environments can be employed to achieve an even faster set of test results.

Given that cyber weapons can selectively deploy a variety of payloads, it may also be possible to fully automatically generate cyber weapons. This level of evolutionary computation can lead to a variety of outcomes including ones that could counter a counter-attack or even completely change the direction of the original "cyber attack," causing confusion at the repair level.

Model/Object Dichotomy Collapse

One of the more fascinating ideas behind the comparison of cyber and kinetic weapons is that in cyber, the weapon's "design" is now identical to the actual "weapon/object." That means, one also no longer needs to use the lengthy process of (1) design to (2) manufacture to (3) shipping to (4) testing and potential (5) alterations of the design all the while disclosing the weapon to a wider body of individuals.

Having the design of a kinetic weapon or the chemical make-up of a weapon does not mean you can make it. In fact, you may be a long way from being able to put any of that information to use. However, having the design of software brings you much closer to having the actual software, and in the hands of an experienced programmer, they are effectively the same.

Companies and agencies have filed classified patents via trusted and cleared patent agents or patent attorneys to protect their inventions.[22] They may then employ contracted trusted and cleared

22 *"Top Secret Patents"* (Collins, 2009), *"UK keeps three times as many patents secret as the US"* (Marks, 2010), and *"143 New Patents That Won't See the Light of Day"* (Marshall, 2011) all reference classified patents.

private organizations to manufacture and ship the inventions to the government for use. This involves many external trusted hands on the invention and product while attempting to uphold supply chain integrity.

Cyber weapons, on the other hand, can be treated strictly as trade secrets, similar to Google's search algorithm or Coca-Cola's beverage recipe, and only employ a select group of trusted and cleared individuals within the chain of command. This model limits exposure of the design, which, as stated above, is the object.[23]

In order to manage the proliferation of cyber weapons, access to the design of cyber weapons will need to be blocked to keep them out of harmful hands. Once a cyber mercenary gets their hands on a cyber weapon, they can easily deploy it with very little or possibly no training. Because the theft or copy of a cyber weapon is easily achieved, as is the ability to distribute it (thereby arming other cyber mercenaries), a very low barrier to entry exists in cyber operations.[24]

Due to this low barrier, nations who would likely never be able to arm themselves with appropriate kinetic capability to stand up against heavily-armed nations or organizations would now be in a position to be considered as a qualified threat in cyber.

Efforts to slow the trade or exchange of cyber weapons may prove to be very difficult to manage via common embargoes and sanctions. New methods of such controls will need to be conceived and put into action. A step further to this problem is the concept of cyber disarmament, which seems nearly unrealistic. With traditional kinetic weapons, the manufacturing process requires raw materials and factories, along with lead times. For cyber weapons, we can effectively avoid this delay and move directly to the object at nearly zero cost, or free replication.

---

23 See World Intellectual Property Office (WIPO), "*Patents or Trade Secrets?*" web page. (World Intellectual Property Office (WIPO), 2017)

24 Solon describes the case where a group called "*Shadow Brokers*" were offering what seemed to be part of an NSA toolset (or state sponsored cyber weapons) operated by a group known as the "*Equation Group*." (Solon, 2016) The starting bid for the package set of tools was 1B Bitcoins or approximately $580M USD.

Expanding on free replication, we face new challenges compared to the kinetic world:

1. Piracy: Traditional software or data piracy is managed through lawsuits and other legal frameworks; however, when it comes to cyber weapons, lawsuits are not a viable solution;
2. Replication on the fly: As one fires a cyber weapon, there is a possibility of the weapon replicating itself as it travels towards its target; and
3. Rearmament: Cyber no longer requires the rearmament of threat actors. It may simply require new compute capability.

Figure 3. Mind Map of Cyber Operations - Model/Object Dichotomy Collapse

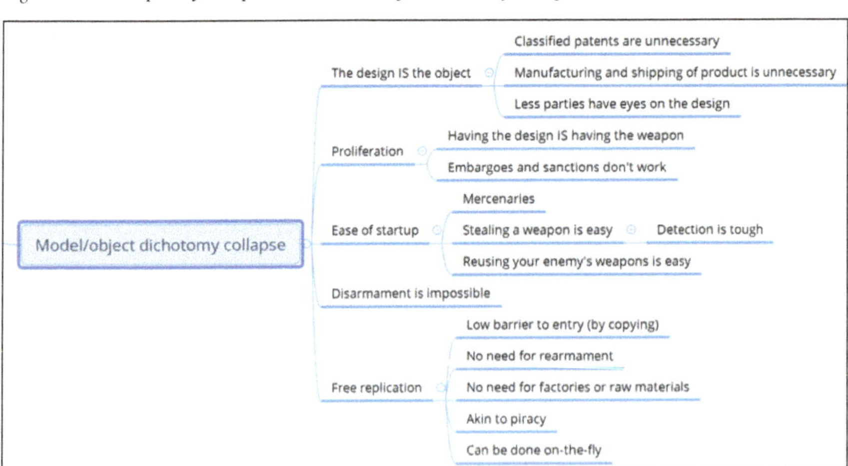

Uniquely to cyber, when a primary source sends a cyber payload to its primary target, the primary target may be capable of realizing the payload and reusing it on an alternate or secondary target. This concept and capability of one's enemy reusing ammunition or weapons after they have been fired is extremely challenging to manage. If the code is in *any way* attributable to the primary source, then the alternate or secondary target may consider the primary source as the true attacker. Possibly worse, any target may be in a position to review the payload, learn from it and thereby increase their maturity, and create a more damaging cyber weapon, all the while making it look as though the original attacker created it. These attribution challenges should force nation states to strongly consider the risks of cyber weapons prior to their use.

An example of this behavior is the Stuxnet attack on Iran's uranium-enriching centrifuges. Although not directly mentioned within the leaked classified document, USA's National Security Agency (NSA) and the United Kingdom's (UK) Government Communications Headquarters (GCHQ) refer to "*Western activities against Iran's nuclear sector.*" The document further states that Iran has likely learned from attacks such as Flame, Duqu, Wiper, and Stuxnet and "*has demonstrated a clear ability to learn from the capabilities and actions of others.*" (Intercept, The, 2013, pp. 1-2)

Intelligence supports that Iran replicated the techniques of the attacks in their own attack directed at Saudi Aramco via Shamoon, which is believed to mimic Wiper. Nevertheless, a group called "*Cutting Sword of Justice*" (Wikipedia, 2017) took credit for the Saudi Aramco attack. Strangely, a Wiper-inspired variant with unknown attribution overwrote critical portions of the hard drives of various computers impacting Sony and various banks and other companies in South Korea. Although NSA and GCHQ could not confirm the attribution of the Shamoon attack on Saudi Aramco or any future attack, "*we cannot rule out the possibility of such an attack, especially in the face of increased international pressure on the regime.*" (Intercept, The, 2013)

Speed of Light

Another key difference between kinetic and cyber weapons rests in the fact that cyber weapons are purely digital as electrons or photons over wire or fiber networks. That means traditional means of early warning systems are no longer effective (example: RADAR) in detecting something also traveling at (or close to) the speed of light. Detection of malicious activity using various log files is not very effective.[25] Real time methods of malicious detection of activity and data exist; however, in many cases they require a significant amount of effort to develop and maintain current as the payload is hidden among expected good data.[26] This may bring forward the interest to possibly slow the data flow down in order to properly analyze the traffic as it approaches a network. Although possible, this is generally not the intended outcome of high-speed networks and intercommunications.

It is also interesting to note that when a cyber weapon payload arrives, the package may take some time to realize this and may not be ready or triggered to deploy immediately. This delay generally offers defensive systems some time to detect and react accordingly.

Figure 4. Mind Map of Cyber Operations - Speed of Light

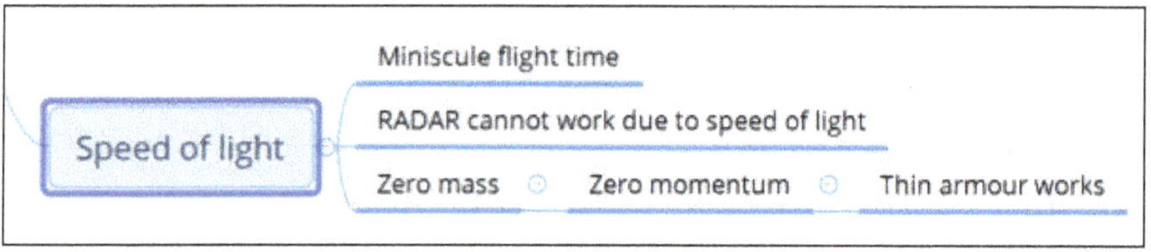

The good news is, since the cyber weapon has zero mass and therefore zero momentum, it does not take much armor to defend. It does, however, take very smart armor or active defence systems. Today's detection techniques and software tend to employ common tools with well-known detection and alert mechanism. Threat-hunting tools tend to use historical data to assign a baseline then compare any new activity against the baseline. Although some of

25 Mandiant states "*...attackers still had a free rein in breached environments [for] far too long before being detected—a median of 205 days in 2014 vs 229 days in 2013.*" (Mandiant, A FireEye Company, 2015, p. 1)

26 Filkins, page 5 paragraph 4, section "*Teaching Machines to Identify Threats*" states: "*Specific domain knowledge related to security...A data scientist must apply security domain knowledge to identify primary and secondary sources of data, determine how to clean and transform acquired data and select the best ML analytical method or algorithm for the problem at hand.*" (Filkins, 2015, p. 5)

this can be automated, the response from the automation software tends to result in several false positives.[27]

To minimize false positives, hands-on analysis of the data with computer-assisted threat-hunting techniques can be employed. Once again, the result of this effort tends to report its findings much later than one would need to defend against a cyber payload arriving immediately. It should be noted that without proper cyber hygiene and strong IT Security maturity, the ability to apply threat-hunting techniques tend to be next to impractical and unachievable.

Dynamic Multidimensional Space

In traditional kinetic war, the theater of operation is normally a physical space. Generally, theaters of operations are located in unpopulated areas for many reasons, including limiting collateral damage. It can be very obvious when one approaches the theater of operation using vessels or fleets of vehicles. This approach grants observing nations some time to decide how best to react.

Distance based artillery weapons take time to approach the theater of operation while in flight. From a defensive approach, the reaction time available to counter these while in mid-flight can be considered beneficial. Methods to track launched artillery, realize the trajectory, calculate the intended location of impact, and even estimate the blast yield based on recognition of the type of artillery are tools that are used by nation states who may not even be directly implicated by the activities or directly involved in the theater of operation.

In cyber, the theater of operations does not mean much. Traditional notions of trajectory, location of impact, and blast yield estimation are completely changed if not gone. It is challenging to detect a launched cyber weapon and even more challenging to track it mid-flight. These challenges may seem similar to those

27 Lee et al. discuss the increased false positives when applying Security Operations Centre (SOC) based hunting. (Lee & Lee, 2017, pp. 8-9)

described in the section "Speed of Light," but they pose another difficulty when it comes to the theater of operation. Cyber allows for a completely dynamic theater of operation, one where collateral damage is very easy.

In cyber space, determining "what is connected to what" is complex; therefore, determination of the damage impact and collateral damage of a cyber weapon is unpredictable. To leverage the unpredictability, the attacker may decide to utilize multiple sources within the cyber space to launch their "cyber attack" making it difficult to determine the true source. The adversary could also vary the frequency of the attacks along with the sources, causing further confusion at the defensive end. Adversaries using virtual hosts to launch their attacks could easily change additional variables of the situation by destroying the virtual images of the various sources.

When looking at the challenge from a corporate or individual perspective, the idea of investigation is essentially akin to self-policing. When a civilian is faced with a break-and-enter at an office or home, normally the primary action is to call the police. The authorities manage the incident by writing a report, performing an investigation, and, much more to hopefully solve the criminal case and under legal governance, bring justice for the injured party.

With cyber, civilians tend to take on the investigative effort themselves. This behavior is similar to tampering with evidence. IT staff spend time combing through log files and looking up Domain Name System (DNS) records to identify domain owners and in some cases directly contact them. With the absence of police capability to support cyber crimes, civilians are essentially left on their own to police themselves. In areas where police are involved, a lack of uniformity between municipal, provincial/state, and federal levels constructs a "craft" knowledge economy way of cyber policing.

Scope of Impact

Cyber weapons have a wide range of initial impact with various subordinate and sometimes unpredictable impacts. When considering the purely digital impacts, one can easily notice the similarity of behavior with traditional espionage. Many news reports describe the negative cyber events as attacks; however, if data has not been lost permanently and is simply duplicated elsewhere (for example), is this behavior an attack or a type of espionage?

At the World Wide Cyber Threats Hearing on September 10, 2015, now former US NSA Director Admiral Michael Rogers stated that

> terminology and lexicon is very important...and attack and act of war...it's not necessarily in every case how I would characterize the activity that I see. (Permanent Select Committee on Intelligence, 2015, pp. 51m20s-52m02s)

Immediately thereafter, now former US Director of National Intelligence James Clapper reinforced the comment by stating:

> Just using the OPM breach is a case and point. That really, although it's been characterized by some loosely as an attack, it really wasn't, since it was an[sic] entirely passive and it didn't result in destruction or any of those kinds of effects, so that the distinction you pointed out, and thank you for doing that, is quite important. (Permanent Select Committee on Intelligence, 2015, pp. 51m20s-52m02s)

In the traditional world of espionage, capturing information or even altering information or public perception is expected criminal behavior, even during times of peace[28]. The similar behavior in cyber

28 Libicki describes *"...espionage by countries has been treated as acceptable state behaviour...This understanding has been carried over into cyberspace."* (Libicki, 2017, p. 2)

could be called "cyber espionage" instead of "cyber attack" and result in a criminal case versus a nation state's cyber or kinetic response.

If one were to take such a position, then the alteration of data, such as the WannaCrypt (Wikipedia, 2017) or Petya (Wikipedia, 2017) ransomware, or deletion of data, such as the Shamoon (Wikipedia, 2017) virus, may also fall into the category of espionage and sabotage. The loss of data should be expected, as it could happen due to hardware or software failures at any time. Therefore, other means of data backup and restore capabilities should exist. When other means of data protection do not exist, one could interpret this as negligence. Hence, if data is destroyed via cyber behavior, then one should be able to restore the data using well-defined and mature Information Technology management processes.

Taking this theory further, leaving misinformation behind and/or highlighting specific information are also methods of espionage. If one were to leave a package in a data store or deposit a package via any means into a computer network resource, could this behavior be considered espionage? In his paper, Singer compares land mines and Improvised Explosive Devices (IEDs) to autonomous cyber weapons as it applies to the direct participation of hostiles, describing the challenges faced with respect to the Law of Armed Conflict (LOAC)[29].

Tit for tat and Game Theory: Typically, a kinetic attack generally expects a kinetic response. So, would then a "cyber attack" expect a cyber response?[30] When it is difficult to prove attribution, where should a nation state direct its cyber or kinetic response? When the source of cyber behavior can be misrepresented as almost anyone's computer, such as a specific civilian's, how is a nation state able to identify the true source? Would it then attack a civilian? If a precedence is set to attack civilians based on cyber behaviors, then

29 Singer describes *"Often the situation is compared to a civilian placing a mine or an IED, who is regarded as not directly participating after its return, completing the action (the revolving door problem)…"* (Singer, 2017, p. 9)

30 On the future of cyber, on August 14, 2017, advance to 49m 46s to 50m 03s of Lt. Gen. Stewart's full speech at DoDIIS 2017: *"I want to be able to hunt for and isolate that malicious strange behaviour. Once we've identified and isolated that malware, I want to analyze it, re-engineer it, and prepare to deploy it against the very same adversary who sought to use it against us."* (Stewart, 2017)

nation states setting this precedence should expect reciprocal actions when the roles are reversed. The challenge of attribution plagues the entire scope of cyber.

Levels of attribution are also a measure of traditional espionage. A nation state may choose to directly engage with another nation's individual(s) to perform a criminal act, such as extract information, affect change regarding a specific policy, or infiltration. Equally, the same nation state may decide to perform the same task using a level of attribution that distances them from the actual criminal act, making it difficult to definitively identify state level involvement in court or even at an international community level.

Even if a nation state follows laws and does not target civilians or civilian services when performing cyber activities, their adversaries may not choose to behave the same with their response. According to James Clapper,

> The big take away for me, is that unless you are very confident in your cyber defenses, it's almost pointless to talk about cyber attacks. The very essence of offense is, you have to have a good defense, ... And what complicates it further is, we in the U.S., we have an inclination to be very precise, very limited, very surgical, legalistic. You cannot be assured that the adversary is going to be similarly precise and surgical and legalistic. So if you attack them, you have [to] anticipate a probably much ... greater retaliation as a result.[31]

With respect to the scope of impact, his thoughts shed light on another challenge with cyber activities. Countermeasures may not always be the answer as they may simply result in an escalation of countermeasures.

---

31 Former US Director of National Intelligence, James Clapper, made the statement during his keynote on September 28, 2017 at the ICF CyberSci Symposium in Fairfax Virginia, USA. Waterman summarizes the keynote delivered by now former US Director of National Intelligence James Clapper including the quote. (Waterman, 2017)

Thus far, the following examples fall into the purely digital form of attack, which may likely be better described as espionage, and their impact to the data's *confidentiality, integrity,* and *availability* (CIA):

1. Consumption of computational resources, impacting data availability;
2. Blocking of services, impacting data availability;
3. Copy (or theft) of information, impacting data confidentiality;
4. Alteration of data, impacting data integrity;
5. Deletion (or destruction) of data, impacting availability; and
6. Leaving data behind, impacting data integrity.

Each of these cyber espionage behaviors are most likely criminal behavior, and similar to traditional espionage.

When it comes to assessing the emergence of cyber behavior into the non-digital space, the impact is not as clear. To remove some of the obvious items from this area, let us ignore systems that are air-gapped by one or more levels. Breaches of air-gapped systems tend to use more than just cyber and thereby likely fall under traditional espionage behavior. That tends to leave non-critical systems behind. Systems in the cloud, fog (example: Wi-Fi space), or mist (example: Bluetooth space) would once again each be challenged by attribution.[32]

An example of cloud implications affecting a large audience could be a government-provided nationwide social service program. In this example, users log into a secure platform online and enter or retrieve their own personal data regarding their social program. Destroying this platform via criminal cyber behavior (impacting data availability) would likely result in a data restore-and-repair plus the closing of the vector of attack. Altering of large amounts of data (impacting data integrity) would likely be noticed with proper IT management and not allowed. Altering of data specific to one or

32 Preden's whitepaper describes the concepts of cloud, fog, and mist computing. While lecturing at the Tallin University in January 2015, Dr. Preden created the term "*mist computing.*" (Preden, 2016)

few individuals (impacting data integrity) may go unnoticed for a while, but the impact may be small and easily repaired with proper IT management and good cyber hygiene. Harvesting of all data (impacting confidentiality) should not be possible when common practices and proper system architecture are applied, and if it were to be possible, only portions of the data would be exposed resulting in a less or unusable incomplete data set. Either way, the entire cyber behavior is criminal and likely challenging to definitively decide to start a war over.

Figure 5. Mind Map of Cyber Operations – scope of impact

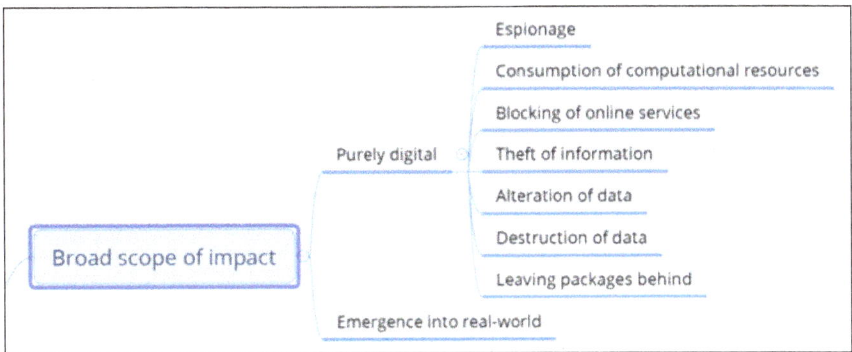

An example of fog implications affecting a large audience could be a deployment of (1) public Wi-Fi access points susceptible to malicious code or (2) illegitimate cellular radio towers, whereby the impact can be quite large, affecting public and private sector users. Normally, if one is communicating at a classified level, these types of public devices cannot be used. Corporate users or unclassified security-conscious users would likely consider the use of Virtual Private Networks (VPNs) for remote access and employ cyber security common practice controls to limit their risk of exposure. Notably, use of public voice networks to convey tactical responses that will be carried out immediately is allowed in many countries. Hence, the interception of such a call is not much of a concern at a nation-state level. Finding the illegitimate cellular radio towers may prove to be challenging but are generally easy to detect by those who are well trained in the art.

Public Wi-Fi access points that are susceptible to malicious code should be updated and maintained with proper IT practices. Any device that is suspect due to potentially-compromised code/design or supply chain integrity should be removed from service and replaced with one that does not carry these risks.[33] Once again, the behavior described is most likely criminal and not one that is causing physical harm or death.

Finally, applying an example to the mist where a large audience is impacted could be a mass exploit of the Bluetooth protocol. Compromised devices could be openly sharing personal information, such as contacts stored in cellular phones to credit card information stored within personal mobile devices. Once again, classified devices generally have wireless protocols and wireless capability physically disabled, so there is no impact to these types of devices. Personal and corporate devices impacted by this type of infection would likely be cross-sharing information with each other (impacting data confidentiality). This type of information exchange would result in the rapid use of one's mobile device data storage.

The infection could be easily stopped at an individual level by disabling the Bluetooth antennae or even completely disconnecting from all networks (also known as airplane-mode on some devices). When considering the case of involuntary sharing of credit cards and if an unauthorized user could misuse that credit card information, the risk is mitigated by the credit card companies who allow only low value transactions via mobile devices. All transactions remain monitored by the credit card company and they hold the right to mark the transaction as potentially or certainly fraudulent, thereby disallowing the transaction and likely assigning the card as compromised

33 Palmer describes the use of hotel WiFi access points to deliver malware (Palmer, 2017)

(impacting availability). In any case, this would result in a criminal case and most likely cause mass interruption, but certainly no direct physical harm or death with any expectation of a nation's kinetic response.

Building on the concept of zero mass and zero momentum from the section on "Speed of Light," is the measure of energy. Kinetic weapons have a quantity of joules or kinetic energy applied and delivered. The energy calculation can be streamlined by using the mass and velocity of the object. A 10 mm bullet (with a mass of 0.015 kg)[34] leaving the muzzle of an MP5/10 submachine gun (with a velocity of 425 m/s)[35] results in approximately 1,355 joules of kinetic energy per bullet (using the formulas). Applying the same logic, with 0 g of mass and velocity equivalent to the speed of light, cyber weapons deliver 0 joules of kinetic energy. Therefore, the impact in the physical space is effectively not possible without its first being converted into real-world (non-cyber) kinetic energy.

From Domain of War to Tradecraft

When we sit back and look at the history of events of any situation, we tend to see things differently as hindsight is 20/20. When looking back at cyber, many are simply astonished with how far we have come. When reviewing the historical and future behavior of cyber and comparing it to non-cyber activity, the relationship with espionage surfaces.

During the Q&A after his speech on August 8, 2017 at the annual Space and Missile Defense Symposium in Huntsville, Alabama, General John Hyten, commander of US Strategic Command (USSTRATCOM), stated:

34 Wikipedia article on "*10mm auto*" where the largest mass described is 15 grams (Wikipedia, 2017)

35 Wikipedia article on "*Heckler and Koch MP5*" where the MP5/10 muzzle velocity is 425 m/s using 10mm calibre rounds (Wikipedia, 2017)

> There's no such thing as war in cyberspace. There's just war. We have to figure out how to defeat our adversaries, not to defeat the domains where they operate. (Hyten, 2017)

Hyten's point was to state that the US will defend and deter an adversary with any means necessary. Nevertheless, a nation state may still decide to declare cyber as a Domain of Warfare, bringing with it its own set of challenges. From the difficulty to definitively apply attribution, to the direct participation of hostiles with autonomous cyber weapons (Singer, 2017), to the inability to truly cause physical harm similar to a kinetic weapon, the cyber domain is more consistent with espionage and sabotage.

A key difference between traditional and cyber espionage or sabotage is the deterrence with risk of legal punishment. When one performs traditional espionage or sabotage in a foreign country, they are considered a hero in their own country and a criminal in the opposing. If they commit acts of treason against their own country, it is generally the opposite result.[36] Either way, aside from the legal consequences, there may be a risk to one's self, family, and possibly even friends' safety. Because traditional espionage involves the actor to physically enter another country or leave their own, there is a higher risk of being arrested, commonly attracting little to no media attention.

Under cyber, the behavior may be compared to a dog with a loud bark but almost no bite. Attribution aside, international political and civilian pressures drive sympathies for the crimes, commonly due to higher media attention.[37] Furthermore, you need to wait for

---

36 As retired CIA counterintelligence analysts, Sandra and Jeanne describe CIA counterintelligence officer Aldrich Ames as a mole and traitor to the US. Ames provided US intelligence information to the Soviet Union that resulted in the deaths of more than ten Soviet intelligence officers who spied for the US. Sandra described her sympathy for the Soviets and their family members and attributed their deaths to Ames. (Grimes & Vertfeuille, 2012)

37 BBC News describes sympathy offered by American Civil Liberties Union (ACLU) and Amnesty International as they campaigned to have Snowden pardoned. The article also highlights the 2016 "Snowden" film's director's comments (Oliver Stone) describing the US government's activities as

the criminal to enter your country[38] or request extradition, only to be challenged by sentencing guidelines.[39]

What may be plausible is that the idea of full kinetic war is not the goal. This is likely because engaging in full kinetic war with capable adversaries tends to result in a level of mutually-assured destruction. This level of deterrence is obvious with nations capable of executing nuclear options.

So, then, what if the meaning of war is no longer only kinetic? Today's conflicts, commonly called wars, seem to occur for ideological, political, and economic factors. These conflicts tend to be fought less with kinetic options and more with sanctions, embargoes, political pressures, policies, and most certainly various methods of espionage. This essentially means nation states are at some level of war, or constant conflict, all the time. If constant conflict is today's new model of living, then it is very likely that the tradecraft of the cyber domain is something that may be useful as espionage and sabotage tools to assist in warfare.

The idea that cyber is shape-shifting may come from the wide interpretations and expectations of cyber. It is our opinion that cyber effectiveness is highly dependent on the target's cyber hygiene and the adversary's ability to exploit or infiltrate it. It is similar to the idea that one's ability to avoid HUMINT exploitation and infiltration is highly dependent on how they protect and live their lives. Poor life decisions can likely lead you to be considered a better and/or easier target.

To reiterate, the purpose of this paper is not to apply a definitive outcome or decision. Rather, it brings a taxonomy to bear—highlighting the differences between cyber and traditional

---

"illegal," thereby indicating sympathetic support for Snowden. (BBC News, 2016)

38 Perez reported that a Chinese national was arrested in relation to the 2015 US OPM breach when he entered the US attempting to attend a conference. The charges are related to the creation of the Sakura malware, leaving the reader to assess if other arrests could be made regarding the use of the malware regarding the 2015 US OPM breach. (Perez, 2017)

39 Williams describes how judges are struggling with the sentencing of cyber crimes. (Williams, 2016)

warfare domains—and provokes the reader with an alternative view of the key concepts, which may change the way we perceive and consider cyber operations. What may be the impact of calling every cyber event a "cyber attack?" The continued misuse of the term "cyber attacks" can easily lead many to incorrectly identify the situation(s) as a legally-defined "attack," thereby applying public and political pressure for a nation state to respond with possibly poor decisions of great consequence. Alternatively, "crying wolf" may well desensitize the public to the point that actual attacks unfold without adequate responses.

References

Australia Computer Emergency Response Team (CERT). (n.d.). Retrieved November 6, 2017, from https://www.cert.gov.au/

BBC News. (2016, September 12). *Edward Snowden: ACLU and Amnesty seek presidential pardon*. (BBC News, US & Canada) Retrieved November 6, 2017, from BBC News, US & Canada: http://www.bbc.com/news/world-us-canada-37341804

Benzmüller, R. (2017, April 10). *Malware trends 2017*. (G DATA Software) Retrieved November 6, 2017, from G DATA Software: https://www.gdatasoftware.com/blog/2017/04/29666-malware-trends-2017

Boyle, D. (2014, December 27). *Hackers release cache of 13,000 passwords and credit cards of PlayStation, Xbox and Amazon users*. (Daily Mail, MailOnline) Retrieved November 6, 2017, from Daily Mail, MailOnline: http://www.dailymail.co.uk/news/article-2888339/Hackers-release-cache-13-000-passwords-credit-cards-Playstation-Xbox-Amazon-users.html

Calabresi, M. (2017, May 18). *Inside Russia's Social Media War on America*. (Time Magazine) Retrieved November 6, 2017, from Time Magazine: http://time.com/4783932/inside-russia-social-media-war-america/

Carnegie Mellon University CERT®. (n.d.). *The CERT® Division*. (Software Engineering Institute, Carnegie Mellon University) Retrieved November 6, 2017, from http://www.cert.org/

CERT®. (2000, May 9). *Advisory CA-2000-04, Love Letter Worm*. (Software Engineering Institute, Carnegie Mellon University) Retrieved November 6, 2017, from https://www.cert.org/historical/advisories/CA-2000-04.cfm

CERT®. (2001, September 25). *Advisory CA-2001-26, Nimda Worm*. (Software Engineering Institute, Carnegie Mellon University) Retrieved November 6, 2017, from https://www.cert.org/historical/advisories/CA-2001-26.cfm

CERT®. (2003, January 27). *Advisory CA-2003-04, MS-SQL Server Worm (SQLSlammer)*. (Software Engineering Institute, Carnegie Mellon University) Retrieved November 6, 2017, from https://www.cert.org/historical/advisories/CA-2003-04.cfm

CERT®. (2004, January 30). *Incident IN-2004-01, W32/Novarg.A Virus (Mydoom)*. (Software Engineering Institute, Carnegie Mellon University) Retrieved November 6, 2017, from http://www.cert.org/historical/incident_notes/IN-2004-01.cfm

Clapper, J., Lettre, M., & Rogers, M. (2017). *Joint Statement for the Record to the Senate Armed Services Committee, Foreign Cyber Threats to the United States.* United States Senate Committee on Armed Services. Retrieved November 6, 2017, from https://www.armed-services.senate.gov/imo/media/doc/Clapper-Lettre-Rogers_01-05-16.pdf

CMMI Institute. (n.d.). *What is Capability Maturity Model Integration (CMMI)?* (CMMI Institute) Retrieved November 6, 2016, from CMMI Institute: http://cmmiinstitute.com/capability-maturity-model-integration/

Collins, B. (2009, July 10). *Top Secret Patents.* (Inventors Digest Magazine) Retrieved November 6, 2017, from Inventors Digest Magazine: https://www.inventorsdigest.com/articles/top-secret-patents

Filkins, B. (2015). *The Expanding Role of Data Analytics in Threat Detection.* (SANS Institute) Retrieved November 6, 2017, from SANS Institute: https://www.sans.org/reading-room/whitepapers/analyst/expanding-role-data-analytics-threat-detection-36362

Grimes, S., & Vertfeuille, J. (2012). *Circle of Treason: A CIA Account of Traitor Aldrich Ames and the Men He Betrayed.* Annapolis, MD, USA: Naval Institute Press.

House of Commons of Canada. (2017, June 20). *Bill C-59: An Act respecting national security matters.* (Parliament of Canada) Retrieved November 6, 2017, from House of Commons of Canada: http://www.parl.ca/DocumentViewer/en/42-1/bill/C-59/first-reading#enH2829

Hyten, J. (2017, August 8). *Transcript of Gen. John Hyten's speech from the Space and Missile Defense Symposium.* Retrieved November 6, 2017, from US Strategic Command: http://www.stratcom.mil/Media/Speeches/Article/1274339/space-and-missile-defense-symposium/

Intercept, The. (2013, April 12). *Iran—Current Topics, Interaction with GCHQ.* Retrieved November 6, 2017, from https://theintercept.com/document/2015/02/10/iran-current-topics-interaction-gchq/

Lee, R., & Lee, R. M. (2017). *The Hunter Strikes Back: The SANS 2017 Threat Hunting Survey.* SANS Institute. Retrieved from https://www.sans.org/reading-room/whitepapers/analyst/hunter-strikes-back-2017-threat-hunting-survey-37760

Libicki, M. (2017). The Coming of Cyber Espionage Norms. *2017 9th International Conference on Cyber Conflict: Defending the Core.* Tallinn, Estonia: NATO CCD COE Publications, Tallinn. Retrieved from https://ccdcoe.org/sites/default/files/multimedia/pdf/Art%2001%20The%20Coming%20of%20Cyber%20Espionage%20Norms.pdf

Lyden, J. (2011, January 25). *Bastard child of SpyEye/ZeuS merger appears online: Malware lovechild monst(e)r/demon*. (The Register) Retrieved November 6, 2017, from The Register: https://www.theregister.co.uk/2011/01/25/spyeye_zeus_merger/

Mandiant, A FireEye Company. (2015). *M-Trends 2015: A view from the front lines.* Mandiant. Retrieved from http://www2.fireeye.com/rs/fireye/images/rpt-m-trends-2015.pdf

Marks, P. (2010, March 23). *UK keeps three times as many patents secret as the US*. (New Scientist) Retrieved November 6, 2017, from New Scientist: https://www.newscientist.com/article/dn18691-uk-keeps-three-times-as-many-patents-secret-as-the-us

Marshall, E. (2011, October 21). *143 New Patents That Won't See the Light of Day*. (Science Magazine) Retrieved November 6, 2017, from Science Magazine: http://www.sciencemag.org/news/2011/10/143-new-patents-wont-see-light-day

McGuinness, D. (2017, April 27). *How a cyber attack transformed Estonia*. (BBC News) Retrieved November 6, 2017, from BBC News: http://www.bbc.com/news/39655415

Middleton, B. (2017). *A History of Cyber Security Attacks: 1980 to Present.* New York: Auerbach Publications.

Minárik, T. (2016, July 21). *NATO Recognises Cyberspace as a 'Domain of Operations' at Warsaw Summit*. (INCYDER News, NATO CCDCOE) Retrieved November 6, 2017, from INCYDER News, NATO CCDCOE: https://ccdcoe.org/nato-recognises-cyberspace-domain-operations-warsaw-summit.html

Mooallem, J. (2013, August 31). *Squirrel Power!* (Sunday Review, The New York Times) Retrieved November 6, 2017, from Sunday Review, The New York Times: http://www.nytimes.com/2013/09/01/opinion/sunday/squirrel-power.html

Mueller, P., & Yadegari, B. (2012). *The Stuxnet Worm.* Tucson, AZ, USA: Department of Computer Science, University of Arizona. Retrieved from https://www2.cs.arizona.edu/~collberg/Teaching/466-566/2012/Resources/presentations/2012/topic9-final/report.pdf

NATO Communications and Information Agency. (2016, July 1). *NATO expands cyber defence coverage*. Retrieved November 6, 2017, from https://www.ncia.nato.int/NewsRoom/Pages/160701_NATO-expands-cyber-defence-coverage.aspx

NATO Review Magazine. (2013, June 17). *Cyber - the good, the bad and the bug-free: The history of cyber attacks - a timeline*. (NATO Review Magazine) Retrieved November 6, 2017, from http://www.nato.int/docu/review/2013/Cyber/timeline/EN/

Office of the Director of National Intelligence (ODNI). (2017). *ICA 2017-01D - Background to "Assessing Russian Activities and Intentions in Recent U.S. Elections:" The Analytic Process and Cyber Incident Attribution.* National Intelligence Council, Intelligence Community Assessment. Retrieved November 6, 2017, from https://www.dni.gov/files/documents/ICA_2017_01.pdf

Palmer, D. (2017, July 20). *Hackers are using hotel Wi-Fi to spy on guests, steal data*. (ZDNet, Cyberwar and the Future of Cybersecurity) Retrieved November 6, 2017, from ZDNet: http://www.zdnet.com/article/hackers-are-using-hotel-wi-fi-to-spy-on-guests-steal-data/

Perez, E. (2017, August 24). *FBI arrests Chinese national connected to malware used in OPM data breach*. (CNN Politics) Retrieved November 6, 2017, from CNN Politics: http://www.cnn.com/2017/08/24/politics/fbi-arrests-chinese-national-in-opm-data-breach/index.html

Permanent Select Committee on Intelligence. (2015, September 10). *World Wide Cyber Threats Hearing*. Retrieved November 6, 2017, from YouTube: https://www.youtube.com/watch?v=Q3aGoCtZbU4

Preden, J. S. (2016, June 22). *Evolution of Mist Computing from Fog and Cloud Computing*. Retrieved November 6, 2017, from Thinnect Inc.: http://www.thinnect.com/static/2016/08/cloud-fog-mist-computing-062216.pdf

Public Safety Canada. (2016, May 04). *Canada Cyber Incident Response Centre (CCIRC)*. Retrieved November 6, 2017, from Public Safety Canada: https://www.publicsafety.gc.ca/cnt/ntnl-scrt/cbr-scrt/ccirc-ccric-en.aspx

Rid, T., & McBurney, P. (2012, February 1). Cyber-Weapons. *The RUSI Journal, 157*(1), 6-13.

Roberts, J. J. (2017, February 15). *Hackers Breach Dozens of Universities and Government Agencies, Report Says*. (Fortune Magazine, Cyber Security) Retrieved November 6, 2017, from Fortune Magazine: http://fortune.com/2017/02/15/data-breach-recorded-future/

Schmitt, M. N. (2012). 'Attack' as a Term of Art in International Law: The Cyber Operations Context. *2012 4th International Conference on Cyber Conflict.* Tallinn, Estonia: NATO CCD COE Publications, Tallinn. Retrieved from https://ccdcoe.org/publications/2012proceedings/5_2_Schmitt_AttackAsATermOfArt.pdf

Shane, S. (2017, September 7). *The Fake Americans Russia Created to Influence the Election*. (The New York Times, Politics) Retrieved November 6, 2017, from The New York Times, Politics: https://www.nytimes.com/2017/09/07/us/politics/russia-facebook-twitter-election.html

Shaw, M., & Garlan, D. (1996). *Software Architecture: Perspectives on an Emerging Discipline.* Upper Saddle River, NJ, USA: Prentice-Hall.

Singer, T. (2017). Update to Revolving Door 2.0: The Extension of the Period for Direct Participation in Hostilities Due to Autonomous Cyber Weapons. *2017 9th International Conference on Cyber Conflict: Defending the Core.* Tallinn, Estonia: NATO CCD COE Publications, Tallinn. Retrieved from https://ccdcoe.org/sites/default/files/multimedia/pdf/Art%2008%20The%20Extension%20of%20the%20Period%20for%20Direct%20Participation%20in%20Hostilities%20Due%20to%20Autonomous%20Cyber%20Weapons.pdf

Solis, G. (2016). *The Law of Armed Conflict: International Humanitarian Law in War* (2nd ed.). Cambridge, U.K.: Cambridge University Press.

Solon, O. (2016, August 16). *Hacking group auctions 'cyber weapons' stolen from NSA*. (The Guardian) Retrieved November 6, 2017, from The Guardian: https://www.theguardian.com/technology/2016/aug/16/shadow-brokers-hack-auction-nsa-malware-equation-group

Stewart, V. (2017, August 14). *Lt. Gen. Stewart's remarks at DoDIIS.* Retrieved November 6, 2017, from YouTube: https://www.youtube.com/watch?v=Nm-lVjRjLD4

Symantec. (2007, January 19). *Trojan.Peacomm (Storm)*. (Security Center) Retrieved November 6, 2017, from Symantec: https://www.symantec.com/security_response/writeup.jsp?docid=2007-011917-1403-99

Symantec. (2012, August 8). *Trojan.Bredolab*. (Security Center) Retrieved November 6, 2017, from Symantec: https://www.symantec.com/security_response/writeup.jsp?docid=2009-052907-2436-99

Symantec. (2017). *Ransomware: 5 dos and don'ts.* (Malware, Internet Security Centre, Norton) Retrieved November 6, 2017, from Norton: https://us.norton.com/internetsecurity-malware-ransomware-5-dos-and-donts.html

Trump, D. J. (2017, August 18). *Presidential Memorandum for the Secretary of Defense, Subject: Elevation of U.S. Cyber Command to a Unified Combatant Command.* Retrieved November 6, 2017, from The White House: https://www.whitehouse.gov/the-press-office/2017/08/18/presidential-memorandum-secretary-defense

Uptime Institute, LLC. (2017, July 20). *Myths and Misconceptions Regarding the Uptime Institute's Tier Certification System*. Retrieved November 6, 2017, from https://journal.uptimeinstitute.com/myths-and-misconceptions-regarding-the-uptime-institutes-tier-certification-system/

US-CERT. (2013, January 24). *Alert (TA09-088A): Conficker Worm Targets Microsoft Windows Systems*. Retrieved November 6, 2017, from Department of Homeland Security: https://www.us-cert.gov/ncas/alerts/TA09-088A

US-CERT. (n.d.). *US Computer Emergency Readiness Team (CERT)*. Retrieved November 6, 2017, from Department of Homeland Security: https://www.us-cert.gov

Vaidya, T. (2015, September 1). *2001-2013: Survey and Analysis of Major Cyberattacks.* Retrieved November 6, 2017, from arXiv, Cornell University: http://arxiv.org/pdf/1507.06673

Waterman, S. (2017, October 2). *Clapper: U.S. shelved 'hack backs' due to counterattack fears*. (Cyberscoop, Scoop News Group) Retrieved November 6, 2017, from Scoop News Group: https://www.cyberscoop.com/hack-back-james-clapper-iran-north-korea/

Watson, B. (n.d.). Knowledge Economies and Cyber. *Postgraduate Diploma in Knowledge and Information Systems Management*. Centre for Knowledge Dynamics and Decision Making, Stellenbosch University.

Wikipedia. (2017, July 29). *10mm auto*, 23:53. Retrieved November 6, 2017, from https://en.wikipedia.org/wiki/10mm_Auto

Wikipedia. (2017, September 29). *Equifax*, 11:43. Retrieved November 6, 2017, from https://en.wikipedia.org/wiki/Equifax

Wikipedia. (2017, August 20). *Espionage*, 16:35. Retrieved November 6, 2017, from https://en.wikipedia.org/wiki/Espionage

Wikipedia. (2017, August 22). *Heckler and Koch MP5*, 07:14. Retrieved November 6, 2017, from https://en.wikipedia.org/wiki/Heckler_%26_Koch_MP5

Wikipedia. (2017, September 8). *Office of Personnel Management data breach*, 16:32. Retrieved November 6, 2017, from https://en.wikipedia.org/wiki/Office_of_Personnel_Management_data_breach

Wikipedia. (2017, August 14). *Petya (malware)*, 16:37. Retrieved November 6, 2017, from https://en.wikipedia.org/wiki/Petya_(malware)

Wikipedia. (2017, July 3). *Shamoon*, 17:17. Retrieved November 6, 2017, from https://en.wikipedia.org/wiki/Shamoon

Wikipedia. (2017, August 10). *Timeline of computer viruses and worms*, 07:39. Retrieved November 6, 2017, from https://en.wikipedia.org/wiki/Timeline_of_computer_viruses_and_worms

Wikipedia. (2017, June 21). *Ufa train wreck*, 01:31. Retrieved November 6, 2017, from https://en.wikipedia.org/wiki/Ufa_train_wreck

Wikipedia. (2017, August 18). *WannaCry ransomware attack*, 20:29. Retrieved November 6, 2017, from https://en.wikipedia.org/wiki/WannaCry_ransomware_attack

Williams, K. B. (2016, January 9). *Judges struggle with cyber crime punishment.* (The Hill) Retrieved November 6, 2017, from The Hill: http://thehill.com/policy/cybersecurity/265285-judges-struggle-with-cyber-crime-punishment

World Intellectual Property Office (WIPO). (2017, August 10). *Patents or Trade Secrets?* (World Intellectual Property Office (WIPO)) Retrieved November 6, 2017, from World Intellectual Property Office (WIPO): http://www.wipo.int/sme/en/ip_business/trade_secrets/patent_trade.htm

Zetter, K. (2016, March 6). *Inside the Cunning, Unprecedented Hack of Ukraine's Power Grid.* Retrieved November 6, 2017, from Wired: https://www.wired.com/2016/03/inside-cunning-unprecedented-hack-ukraines-power-grid/

Neal Kushwaha
neal@impendo.com

Bruce W. Watson
bruce@ip-blox.com

5

Major General Stephen G. Fogarty

As presented at the 2017 Civil-Military Symposium
Hosted by the Institute for Leadership and Strategic Studies
University of North Georgia

Good morning Dr. Jacobs, Dr. Wells, Dr. Payne, my ranger buddy Keith Antonia, distinguished guests, colleagues, and members of the faculty and student body here at the University of North Georgia. And I do have to admit that I'm always nervous when I come back on UNG's campus, particularly when I'm invited because I figure there's either a parking fine or a library fine from 1984 that, if you just put a just small amount of interest on, will be so high that I'm going to end up having to sell my house to get out from under.

So, my wife Sharon, who is also an alum, just said, "Just be very careful. Go out the back door as quickly as possible so we can move on." My boss, Admiral Mike Rogers asked me to pass on his best wishes for this inaugural event. They normally don't let me out of my cage, especially when the boss is traveling; he's in the Pacific this week. And, so, I really appreciate the invitation by Keith to participate today.

I'm going to lead off by stating that I don't presume to lecture you this morning; instead, I'm just going to try to share some thoughts on the lessons that we've learned over the last few years. At Fort Gordon, we built the Army's cyber force. We were the only service to actually stand up a Cyber Branch. It was a strategic decision for the army. I think it has given us a tremendous advantage.

We've looked at the models the other services have used, and what this Army cyber force really offers is the ability for a young enlisted soldier—an NCA, a warrant officer or an officer—to enter the cyber corps and actually do that for the remainder of their career. Prior to this point, the decision that General Odierno, former chief of staff of the Army, made a few years ago, you would serve in some other branch; it was kind of like special operations back in the day or aviation branch back in the day. And you did it for a few years, then you go back to your control branch.

And what we found is that wasn't very satisfying; it didn't develop that core of subject matter experts that we understand that we must have to operate successfully in this space. So, I think the Army's leading the way. The other services have taken a different approach, not because they're not clever or smart; they really selected what was best for their service culture. And I'm going to talk actually a lot about culture today because I think, frankly, it's one of our greatest challenges to becoming much more effective, to getting after this in the manner that we need to.

One of the things that I also want to talk about is, we've conducted operations—it's not just in the cyber domain, I think that's coming back to haunt us to a certain extent. I'll talk about our operations in the information environment, because in my opinion, it's much greater than the cyber domain. I was speaking at a conference of the old Common Remotely Operated Weapons (CROWs), Space Electronic Warfare (SEW) tribe and the army and across the joint force just a couple of weeks ago up at the Aberdeen Proving Ground. And one of the questions during the Q & A period was, "You've talked a lot about information warfare operations, but we have a Cyber Command, we have a Cyber Center of Excellence (CCOE), we have a Cyber Branch, a cyber corps. There's a little dissonance there."

And my response was, "We had to start some place." It was that simple. And so, identification of Cyber Branch, the cyber domain,

CCOE, US Cyber Command—those were important decisions. But I think we're very early in this journey; as an intel guy for the last thirty-four years, I would argue we've been doing this for a long time. If you're in the signal community, you're in the communications community, you would argue we've been doing this for a long time. But what I will tell you is, I think we're at the start of the journey. I predict within five years, we might be the United States Information Warfare Operations Command.

Fort Gordon will not be the CCOE; it'll be the Information Warfare Operations Center of Excellence. And you're going to see this convergence, the integration of all of the information related capabilities. So Psychological Operations (PSYOPS), Military Deception (MILDEC), electronic warfare intelligence, communications, and cyber are just some of them (that I'll talk a little bit about later in my presentation) that we've seen adversaries and potential adversaries integrate and actually employ very skillfully.

So, we have to not only match them, but get ahead, and this is very important. I'm going to caveat my remarks with a kind of a standard disclaimer: they reflect my thoughts alone, and they're not the mission of the US Cyber Command's, my boss's, or the Department of Defense's. I'm also going to admit that in my normal life, when I come in in the morning, when I get home at night, I'm kind of programmed for what I see on National Security Agency (NSA) 'net, and the Joint Worldwide Intelligence Communications System (JWICS) 'net, on Super-Net, and on the classified systems. So sometimes for me, it is a fine line, as I'm having a discussion in an open forum.

And, so, in some cases, you're going to ask a question, and I'm going to thank you very much for the question. I'm just not going to be able to answer it or probably answer it to your satisfaction. And so, that's just kind of the way it is, and that is one of the rules of engagement for me. As I said, I do these in more open forums. You're going to no doubt notice I'm not using PowerPoint, which is pretty unusual for a military guy, particularly an army guy. So, I'm going to use my Central Intelligance Agency (CIA) brothers' approach to doing business, and I'm going to use some notes. I think I'm organized, so it won't be a random stream of consciousness, but I'm trying to cover a lot of ground very rapidly.

Because the reality is, this is a big topic. I'm going to try to distill it down into some essential elements. Colonel Grant and I, he's shaking back there right now. He was the director of the battle lab when I was a commander at the CCOE, and he kind of knows how I operate in these environments. So, I'm going to go ahead.

I actually shied away from a white board today. So first of all, it is a pleasure to join you for this symposium today. It's an absolute honor to be asked to share a few thoughts with you this morning. I think the theme for the symposium is particularly relevant. I want to offer my compliments for your efforts to create this event, identify this particular theme and then assemble this very impressive international team of subject-matter experts to discuss the challenges and hopefully offer a few thoughts for the way forward. I want to thank Dr. Wetherington and the panel.

I think you guys have really kicked this off to a great start, and I hope that I can kind of match the operational tempo that you commenced with. I'm not a theorist; I'm a practitioner. And for me, the task is actually fairly simple, fairly straightforward, and that is to win in a complex world. I'm going to leverage every capability to do this. So, I don't limit myself to cyber, don't limit myself to information warfare operations. When we conduct cyberspace operations, it is part of a whole of government, whole of nation, whole of coalition campaign.

And what we find is, if you attempt to try to use cyber in this very discrete kind of a solo approach, I think in most cases you're going to be doomed to failure. Now, as a military officer in the United States (US) Department of Defense (DoD), I'm aware that implies I have a certain bias. I'm not offended by that, but I also find that it's not unique to DoD. But it is in fact common to all tribes active in this space, and this is the academic tribe, the government tribe, the commercial industry tribe, and, in discussions with my foreign partners, I believe this also is not unique to the US.

So, congratulations on a great symposium. The discussions here today such as here with some of the presenters, some of the panel members who've spoken already, they are very consistent with the discussions we have every single day. And I'm not a traditional chief of staff. Admiral Rogers made that very clear when I signed into the command about sixteen months ago. I have a much more operational role than what a traditional chief of staff would have at a combatant command. And so, as I'm talking to you today, I'm not going to do it from an administrative position or from my last job, although some of that will creep in. But it's literally from that tough point of conducting operations, and I'm talking about a full-spectrum cyberspace operations every single day.

This is an evolving environment. It merits careful analysis, responsible discussion. And I think as we start off, one of the things I want to do is talk about scope, because it's very easy to build this very large elephant, and you really struggle to figure out what you're going to grab onto. So, we have a tendency to do this: we tend to grab onto a little piece that becomes our world, and we have a tendency to describe it, I think, in almost dystopian terms. I'm not in that camp. There's a lot of opportunity as well as a challenge.

And depending on where you sit on this cyber issue, you can range somewhere between global warming and nuclear annihilation. I'm somewhere in the middle of that, frankly, and I'll kind of explain my perspective as we work our way through this. And I also want to caution this audience as you're starting to discuss this, as you're starting to really get your hands around it. I think we need to be much more precise in our description of both the threat and the challenges as well as the opportunities.

The Pearl Harbor analogies, I think, contribute to a certain extent to the cyber fatigue that I encounter as I move around and particularly in the Beltway. It is very easy to get into a defensive crouch; you feel you're under assault and I'll talk about that because we are, but what I will tell you is we give back pretty hard also. I'm also going to talk about culture because as an intel guy, I'm the absolute optimist. My job is to break through every defense and get the crown jewels. And some targets are tougher, some targets take longer, they take more work, they take some enabling capabilities.

But if it's that important to me, I'm confident that I will get through. And, see, you contrast that with the signal community, the signal tribes: that's the intel tribe's perspective. If you contrast that with the signal tribes' perspective, they have ultimate faith in the power of the defense. And, so, we're immediately at odds with each other. And what I'll talk about today is it's not binary, I don't believe. It requires a level of collaboration that's hard to achieve within the military, within a service, much less within the inter-agency, the inter-government, the whole of nation, and the whole coalition approach.

The place has some perspective. I think we're at one of the unique periods in history where a single actor—in some cases perhaps an individual actor or an adversary, it could be a nation-state adversary with limited means or investment—can threaten the security of nation states possessing the most modern and capable conventional and nuclear arsenals. And this requires a determined and pragmatic response and urgent effort. The current situation demands clear and effective policy, effective collaboration between the inter-agency, DoD, academia, industry, and our foreign partners' defensive and offensive capabilities, and, most importantly, the will to act.

Frenetic activity process and platitudes create the illusion of progress, but what they really do is mask failure. In this case, failure to adequately understand the threat and take effective and decisive action. So, a couple of thoughts as I kind of wade into this. We're live in the drain today. This is not a future challenge. It is reality. The Army's looking toward the future; in some cases, they're describing this threat as kind of 2025, a 2030 or beyond threat. And as I've talked to the chief, I've talked to the senior leadership of the army, what I've made very clear is, we're seeing these capabilities being employed today against us, and we're employing these capabilities today against our adversaries.

So, this is not something in the future; it is the reality that we're living with today. To kind of give you an idea, when I came in the Army in 1984, we had a single Internet Protocol (IP) device in a brigade—we didn't call it a combat team back then, but the infantry brigade I served in at Fort Campbell with 101st. Today, if you go to that same brigade, it has almost 20,000 devices. So, we're almost entirely dependent in the US DoD on this capability. So that poses significant vulnerability to our ability to wage warfare in the way that we want to.

And I'll tell you, we didn't necessarily pay close attention to it, but our adversaries did, and they are. And I use a little vignette: we have a general officer corps called Capstone that all of our general officers, many of our Five Eyes officers, participate in. And it's really to expose general officers who may have grown up in their service, potentially within a very narrow specialty within a service, to the kind of the joint capabilities that exist, not just of the military but across the inter-government.

And we host a visit by those, along with our interstate counterparts up at Fort Meade about once a month. And the most recent class came through; I was talking to the class, and I start off that discussion like I start off with most of the discussions I have, at least with the military audiences. And I asked them very simply, "What's the most important weapons system? The most important capability that the United States Department of Defense operates twenty-four hours a day, seven days a week?"

And they really struggle. We got, "Well, the soldier, or the sailor, the M4 rifle, whatever." So now you're actually not even shooting close to the target. I said, "But, if I went to Moscow or Beijing and I sat in on their version of the Capstone class and I asked them what is US DoD's most important capability," I believe they would answer without hesitation the Department of Defense Information Networks, because they understand our dependency . . . They understand the threat that that poses to them, and so they have built their modernization programs designed to directly attack that most important capability that we possess.

So, I'll talk a little bit about that. I think the adversary's outmaneuvering us. They operate persistently, effectively, and cheaply, and they don't realize, I think, we have a bias to solve $10 problems with multimillion dollar solutions. And that places us on the losing end of that curve. So, it's like shooting a Patriot missile down a $400 DigiCam helicopter. This is where we're at in this space, and we've got to reverse that trend. We've got to do it very quickly. Operations in this domain, like every other domain, are driven by intelligence. And this is really important.

And this is where we have a lot of our cultural wars. It's lack of understanding of the importance of intelligence, lack of being able to normalize operations in cyberspace. We have a tendency to think of it as something different, and what I'll tell you the fundamental fact is, it is very much like other domains. We use different Tactics, Techniques, & Procedures (TTPs), we use different tools. But our ability to integrate, our ability to place demands on the intelligence community, to place very precise, very persistent demands and then force them to produce is very important.

And like any other operational domain, I have to be able to conduct my own reconnaissance. And, so, we're very active in that space also, but I'm very dependent upon the national intelligence system to provide the intelligence that we require. I don't think you can really be in the game if you don't have offense and defense. and

I'm not talking about every nation. There are barriers. Frankly one of them is the cost.

One of the barriers is political in some nations. But if you don't have it, you're going to have to partner with someone who does. You cannot impose costs on an attacker without offensive capabilities, and, in fact, I don't believe you can establish deterrence without credible defensive and offensive capabilities and the will to employ the offensive capabilities. So, if we rely on the defense as our deterrence solely, we're going to be sitting like the French army was in 1939 behind a very penetrable barrier. And we place ourselves at unacceptable risk.

So, you better have both in your arsenal; and as for your adversaries, they better understand you have game. And you better have the willingness to employ that game because if you don't and they don't, they're going to be all over you. And that is a significant challenge. It is perception. So, when we talk about information operations, information confrontation, part of this is speaking truth to power. Part of this is absolutely having game and employing game. Sometimes you've got to let them know you've done it. And you cannot self-limit yourself to the cyber game.

So, if we get caught in this cyber v. cyber response—and sometimes when we talk about escalation in this domain, that's what we get trapped into—we're going to self-limit. So, we've got to be very careful also again about looking at cyber as information warfare, information operations. It is a piece of it. We must normalize cyber and info warfare operations. Our adversaries are skillfully leveraging all information action-related capabilities, and they are achieving good effect. I'll talk a little bit about that and give some examples in a second.

For those that are operating in this space, you're going to be measured on what you deliver every single day. You cannot have a bad day; that's the bottom line. I'm going to talk about the need for speed. And that's really important. The temporal aspect of this problem is very rarely discussed. So, I'm going to give it I think a little bit of exposure this morning. We have to avoid admiring the problem or falling into the paralysis by analysis conundrum. When we do that, we lose the initiative.

What I will tell you is, we have challenged the Islamic State of Iraq and Siria (ISIS) in this space; we've conducted operations against ISIS in this space. They are very agile. They're not worried about getting it into the Secretary of Defense's read book. They're not worried about getting it into the President's Daily Briefing (PDB). All they're worried about doing is surviving and attacking, and surviving to attack another day. So, they are very, very agile. Technically, we can run circles around them. Our process is part of our problem. It drives us to the $10 million solution to the $10 problem. It allows an enemy who really doesn't have much game decisive advantage. This is a significant problem.

We must not self-limit. Sometimes the best response to a cyber attack is not a cyber response. Technology is critical, but as Dr. Wetherington very eloquently stated, this is about talent and passion at all levels, and it requires a balance of uniformed/civilian, government/civilian, and commercial/government contractors. And I'll tell you, as I testified before committees, we've talked to staffers. Everyone wants to try to limit us to filling our requirements with just military or just DoD uniforms. And what I've said, first of all, it's not realistic, it's not desirable. I want to balance.

I want continued influx of new-talent people who can look at the problem in a new way, and I want to decide who I want for continuity. So, it is the best; it's the most passionate. I've got to give them opportunities. I have to send them to school. I want to send them to school. I want to give them opportunities within industry. I want to see how other people are operating. And I do not want to burn them out because this is very active space. It's very easy to burn out the best. And that's when we're going to lose them. It's not because they don't have something interesting to do, but this literally becomes their lives, and, thank God, we have enough people who are willing to actually commit really their lives to this.

But what we do see is burnout. And again, we're at the start of this, so we've got to flatten that curve out, and it's going to require the entire community to meet our requirements. Cyber is like Intelligence, Surveillance and Reconnaissance (ISR) capabilities. It is very hard to measure, but you can never have enough. And that is contributing to our problem; that is contributing to this sense within the service, at least, of cyber fatigue. How much is enough? How do you measure it?

If I increase the budget by 30% more, what does that buy me? If I decrement you 30%, I mark your budget 30%, do I accept any additional risk? Is it 30% or is it 80% risk? And, so, we're having a real challenge with that. Because that's one of the things I hope that this community will take on, help us describe what that is. And we've worked with industry. They use a different model; they use a different cost analysis. But what I'm telling you is, as a guy who for many, many years worked the ISR piece, how many predators do you date or how many rapers.

What does a pound of ISR amount to? What does it cost? What does it mean? What does it provide for you? We're at the same spot right now in cyber. The challenges you face—technical, legal, and policy—they're daunting, but with one exception, they're not insurmountable. I think our most challenging issue is culture. And we're going to have to get to that; I'm going to talk about that in a second. I want to caveat everything I'm saying in this way: we're going to operate ethically. We are hiring soldiers, civilians, airmen, sailors—building a workforce with character. We're going to operate ethically; we're not going to take the shortcuts.

I'm going to give you an example of someone who is. In some cases, it appears they are operating effectively. I'm going to talk about why I believe they've accepted a lot of risk. We're going to do it within a defined legal framework, and this is non-negotiable.

And then lastly, this is generally for the military audience, but this is commander's business. But some of our commanders are risk averse because they don't understand what we're talking about, and they can't visualize it. So, how can you defend something you can't see? How can you attack with something you can't see, that you don't understand.

And this is one of the cultural challenges: our ability to integrate capabilities where they need to be; and that's our basis. We have to be able to describe it, and we have to be able to create the tools that allow commanders to visualize what we're describing. That's one of the things Colonel Grant, they're working on really hard. That was one of the toughest demands I placed on them at the cyber battle lab, that is, the ability to create cyberspace situational understanding—not awareness, but understanding, and be able to do it in a way where commanders can visualize, they can sense, understand, decide, and act, and they can do it faster than the adversary, because only then will we have decisive advantage in this domain.

So, if the Taliban can turn circles within you, they enjoy decisive advantage. Doesn't matter how much mass, how much capability we can provide, but if they can sense, understand, decide, and act faster than us, they enjoy decisive advantage. And they don't shoot back quite like some of our future adversaries. They don't have the gain that our future adversaries might. This is a significant challenge. Now, why is cyber important? And I'm going to flip to something that you may find interesting.

So, when we look at market capitalization, in this case, global companies ranked by market capitalization. So basically, how much wealth do they have. When we look at this, Apple retains a pole position for the sixth year in a row. This is from March 17, 2017. The top three companies are unchanged compared to last year. Within the top ten, the newest addition was JP Morgan Chase, but the most important point is General Electric (GE) fell out. And when you look at the top ten, eight of the top ten are either technology or financial services company. The only two kind of traditional industries that are in that top ten are Johnson and Johnson and ExxonMobil. So GE actually fell out.

So that should give you an indication of how important that is to our economy, and if that doesn't, US is still increasing its dominance. Fifty-five out of the top one hundred are US companies, and the top ten are all US companies. Europe continues to fall behind in its share of the top one hundred. So, this is why this is so important, not to DoD but to the nation. The reason the adversary is attacking us when it's not for military purposes is for the same reason that Bonnie and Clyde robbed banks. It's where the money's at. It's where the wealth of the nation is at.

And, unfortunately, many of us really struggle to understand that. Now, for DoD, it is. Law is our secret sauce. So, I've talked to you, said our most important mission we conduct 24 hours a day, seven days a week, is Department of Defense Information Network Operations. Why? Because of the way the US goes to war, we are absolutely dependent upon our ability to implement effective mission command, long range precision strike, effective intelligence, surveillance and reconnaissance, just-in-time logistics, medical evacuation (medevac).

Essentially, everything that we want to be able to do on the battlefield is almost completely dependent upon our ability to implement and execute effective Deparment of Defense Information Network (DoDIN) operations. So, it should be no secret that for US Cyber Command, that is mission number one. For DoD, within the Defense Cyber Strategy, that is mission number one. Because even though many of us don't recognize that, our adversaries do; they have built capabilities, again, to directly attack that tremendous advantage.

And so, what DoDIN allows us to do is ensure uninterrupted access to networks, data, and weapon systems in both a congested and a contested operating environment while denying the same to the enemy. So, this is the blend of support, it's the blend of operations, it's the merging of attack on information-related capabilities. This is why electronic warfare is important. That's why offensive capability is so important. Is PSYOPs, Military Information Support Operations (MISO), military deception is so important, because these are operations. They're not support activities, are not just enabling activities.

And I think what we're finding is, our senior commanders are starting to really understand that. Our adversaries understand that; they're moving aggressively to deny this capability. They are increasingly assertive and capable, and they perceive that benefits outweigh the risk. So, they're on the right side of the curve right now. They have not self-limited. They do not view cyber in isolation like many of us; they take a system approach, and they consider cyber as part of a continuum of information warfare, and they're skillfully integrating the full range of information-related capabilities to effectively attack across all domains, from strategic to tactical.

Now, I'm going to give you an example because sometimes people challenge me. I was at the fire center about eighteen months ago; I spoke to a large conference there. I was in the Ukraine about eighteen months ago; it was kind of following up a lot of work that we had done trying to identify what was happening around the globe. And what I told that audience is, what we're watching our adversaries do—in this case the Russians, I'll call it out by name—what the Russians are doing in the Ukraine is, they're skillfully integrating, synchronizing all aspects of information-related capabilities to effectively find, fix, and finish their adversaries.

In this case, the Ukrainians. You should ask, well, how does this happen? What do you mean? Here's actually what they're doing, defined: they're able to detect and geo locate Ukrainian forces to a targetable degree. And they're doing it by social media scraping; they're doing it by traditional intelligence collection; they're doing it by traditional reconnaissance, direct observation. They have a broad use of unmanned aerial systems, and they have the capability of bringing all of that together to detect, identify, and geo locate their adversaries.

And then they fix them by employing electronic warfare and cyber attacks against those forces. And you ask, how can you take something as nonphysical and actually create a physical effect? Because if I can separate you from your mission command, so you don't know what the unit to your left and right are doing, you can't call for supporting fires, you can't call for close air support, you can't call for a medevac, you can't call for resupply. I have now fixed you in position. You have one of two options; you can withdraw, or you can stay in place and you're completely isolated.

And then they vanish with traditional long-range precision fires and combined arms maneuver. Now, that's a tangible example. Let's go to the strategic example, because they did the exact same thing when they attacked Ukrainian energy companies in Kiev. They did the exact same thing; they did their reconnaissance, they were able to find where the vulnerabilities were, they were able to fix, they were able to finish with a cyber capability, and they were able to amplify that with information warfare capabilities.

So that's the range and one very quick snapshot of what the threat is. Now, up until this summer, US forces had not been attacked from the air since the end of the Korean War. And this summer, ISIS, using drones—just commercial drives that you can buy at Costco—they actually were able to employ that capability to do something that no other military has done since the Korean War. So, this is how fast things are changing. And, so, we have foreign militaries, we have criminal actors, we have in some cases state actors who are either employing criminals, or they're engaging in criminal activity themselves.

They are actually robbing banks today to generate hard currency. That's what they're employing against us. So, this is why this conference is so important, why efforts like this are so important, because we must get to a better place, and we have to do this very quickly. Now, the need for speed. How many of you have read the recent M-Trends 17 report from Mandiant FireEye? If you haven't, I encourage you to do that. You can sign onto their website. The only thing you have to give them is your email.

They'll actually send you the link, and then you can open it up, and then you can look there. Not only 17, but you can look at the reports before that. So, they are one of the leading commercial companies in this business. And what they have done is, they've actually analyzed going back, they've really established kind of the longitudinal data to be able to make some, I think, pretty important comparisons over a period of time. So, they said, "Look, congratulations. The average time that an adversary was in your network on the commercial side in 2016;" they round 16 figures for the 17 report. But the average time was reduced by about 50% between 15 and 16.

So, it went from about 156 days that an adversary was in their network having their way with you to about ninety-seven days. But they make a really important point; we have adversaries today that are employing state-like sophistication. And, so, it takes that quality adversary about three days to get credential-level access to your network, and then they really can have their way with you. So, congratulations, you're now detecting them within ninety-seven days. And I wasn't a math major—I actually had to take math 101 here twice. But what that tells me, that's ninety-four days too late.

That's a significant problem. So, it is their ability to be able to get inside of our circle. Earlier, we talked a lot about our patching regimens. Today, it's about ninety days. Why is it ninety days? Because that was always good enough. WannaCry should have woken everyone up, because everyone for the most part was on a ninety-day program. And, guess what: they struck before the ninety-day clock. And, so, they were very, very effective. Now, as we know, many people weren't on the program at all. So, shame on us. We've actually made it very easy for the adversary to kick sand in our face. And they do it over and over and over again. And that's what's most disturbing.

On the DoD side, what I'll tell you is, we have taken this very seriously; we have, and we're working really hard. There were no penetrations of DoDIN by WannaCry. So that's an example. We don't think that was very sophisticated, though. It's kind of a brute force type of capability. What I will tell you though is, we're on similar programs; what we've realized is, we've got to get to the left of that, because the adversary, I think, has figured that out. When we talk to one of the largest industries that was affected by WannaCry, that had a pretty strong relationship with DoD, they were an international company.

They were hit in the UK; it spread back into their network in the US very rapidly. And what they were able to do was isolate it very rapidly. And when we talked to them, they were on a ninety-day program also. And when we talked to them about twenty-four hours later, they were 100% complete. And when we talked to them, they said, "We could have done it in two or three days, but we just frankly didn't have to." So, the adversary understands that.

So, this goes back to this idea that the temporal aspect of this is so important. So, adversaries are reusing malware; they're re-weaponizing it; they're making minor changes in some cases, and they're getting it to almost entirely new capability. And, so, this goes back to this idea that the actor who can sense, understand, decide, and act faster than his adversary enjoys decisive advantage. So, if you're waiting for the next release, watching someone use malware, re-purposing malware, buying malware all to employ against an adversary, and you can make those minor changes that make it effective, and you can employ that faster than we can patch faster, than we can detect, then you enjoy decisive advantage. And that's what we're seeing.

The temporal aspect of this, I can't emphasize enough. Now, what's DoD doing about it? I'm going to give it to you in a couple of big ideas. First, three principal missions for the DoD piece of this. We've got to defend our own network systems and information, aka DoDIN, we have to be prepared to defend the United States and its interests against cyber attacks of significant consequence. We have a lot of discussions with our inter-agency partners, what is the role? We have a really difficult time explaining this to our congressional overseers. Why isn't the DoD defending the nation? Well, 90% of the infrastructure is commercial.

We stay on the DoD side, so think .mil. We have partners, Department of Homeland Security (DHS) in particular deals with .gov. And then there's tremendous risk. And where's the wealth at? It's in the .com. So, it really becomes this idea of a team effort, the collaboration between the tribes that I talked about. And there are a lot of issues, a lot of challenges and I'll hit a couple of those in a second. And then third, is we have to enable our joint force commanders. Fighting ISIS, and we're supporting central command of Special Operations Command, that's one of our job ones. So, make sure that they can use DoDIN to do what they need to do. It's also to be able to take the fight to the adversary.

Those are our three principal tasks at CyberCom and really for DoD. Now, we've got really five strategic goals for our DoD. I'm not going to hit one that's really important to this discussion, which is to build and maintain robust international alliances and partnerships, to deter shared threats and increase international security and stability. You're actually off to a good start on that. For us, it is generally a Five Eyes conversation, not exclusively Five Eyes, but it is a very easy Five Eyes discussion. And what we've realized, at least within Cyber Command within DoD, that's going to have to expand. And I commend you for actually opening up well beyond Five Eyes, because the partners of the future are actually sitting in this audience today.

We've got to work the private sector better, we've got to work the inner agency better. And what I'll tell you is, the rice bowls—I've described some of the tribal warfare within the army between the signal tribe, electronic warfare tribe, information operations tribe, the intel tribe, yada, yada, yada. What I'm telling you right now is, it is very similar in some respects. But the president, actually, one of his first executive orders was actually to jump on this, and so the different government agencies have been hard at work over about the last 180 days actually trying to figure this out, and they'll report back within the next couple of months.

Now, there are some barriers and impediments to success. I've talked about one: culture. One of the things that culture issues will contribute to is a lack of speed; because you're not transparent, you don't trust each other. And so, a kind of speed of trust is very important. I can't get to where I want to be on the left-hand side of the curve unless I can set up the mechanisms that accelerate speed of trust between the significant partners. Authorities are always a challenge, but what I'll tell you is, generally, if we can articulate, if we can clearly define what we're asking for, then we can get the authorities that are required.

This is where another cultural aspect comes in, at least in our system, and that's lawfare. That is where, frankly, the operators in many cases have an idea what they want to do, and then we just pitch it over to the lawyers and let them grind it out. And, again, what happens is, process grinds our response down to where we're shooting behind the target. The enemy is actually employing; they're flat, they're agile, and they're employing the temporal aspect of this battle space. Will to act is vitally important; will to act, write that one down—everyone should probably have it tattooed on their forearm.

Many cases exist where there's a number of process and activity versus decisive action, and we've got to figure that out. I don't have a solution. If anyone does, I will definitely buy you a dinner or breakfast—I guess at Danny's; it's probably a better deal. I'm going to finish right now. We must be faster; we must be more precise; we must be more lethal than our adversaries. That's going to require us to change organization, processes, TTPs, training, and talent management. The future is even closer to partnership, well beyond Five Eyes. Coalition operations must include training and education, capability development, and improved information and data sharing.

And I'm going to expand that definitely to the Inter-agency, because in DoD, we operate at the highest classification level, and many of our partners don't. And again, that induces latency, because now I'm trying to take something at a classified, a very high classified capability, trying to actually make it useful enough to you, in a timely enough manner, and our current process is actually to induce latency. We fall to the wrong side of that speed curve.

It's the same problem with non-Five Eyes partners. We must evolve to an information warfare operations construct quickly; we must evolve, or we're going to be left behind. Talent matters most, that's the most important part of the discussion today, not running out of people who are willing to work this problem at the highest capability. I was in Canada about two weeks ago, on a panel at a conference there. I thought they were getting to as close as you can get to Canadians fighting—and no offense to my partners here. And, literally, they have created a false dilemma.

They were trying to make it a binary decision on, "Hey, we can bring in people; we just have to apply mass against the problems; this kind of has equality all of a sudden . . . " That is true in some aspects of this. On the other hand, we had a PhD who argued "Look, you must have the PHTs that are on keyboard." And, the reality is, you have to have both, and you have to have everything in between. Do not let it devolve into this binary discussion of one or the other; we've got to have it all. And what I will tell you is, we're so off limit now with college degrees.

There are a lot of programs out there, with industries using it all very effectively. It's not that we can't learn from industry, where they're sending people basically to a trade school to learn what they need to accomplish the task that they've been given. It cuts years off of that education cycle. In some cases, that's what these guys are going to do. In many cases, they will actually re-enter the workforce or re-enter college. That's why online degree programs are so beneficial. I push those; they've got to be quality programs, but that also gives someone who may be working at Roy Camp the base level of IT. As they develop their capabilities, they want to move to the next level, then they can get that degree. But then they also have the experience that they bring. So, we're going to be open to that.

And I think Dr. Wetherington hit it, this reciprocal accreditation. I struggled with that when I was a commander in inter-state Georgia, with then Augusta State University. I made a little bit more progress when I was the actual post commander. But what I'll tell you is, we're not where we need to be, and I'll put a plea out to the Georgia Board of Regents. You have tremendous talent at multiple locations across the state, and you're actually self-limiting. They are going to the University of Maryland; they are going to other universities because they're actually willing to recognize your university or their hard work experience.

So, I encourage you, I implore you, reach out to that community, open up your community, your tribe, to that workforce. You'll get a lot more for it. And I've got one last thing, and I'll turn it over. Humans are more important than hardware; quality is better than quantity; Cyber Forces cannot be mass produced; competent cyber operations forces cannot be created after emergencies occur. Most cyber operations require non-cyber assistance, and if you filled in everywhere I said cyber with special operations forces, those are the soft truths, and they are absolutely applicable to this domain.

Our cyber warriors, they are special operations warriors; they're working at different problem sets. They're just as dedicated—the skills, the dedication, is the same. And we actually wrote an article from the CCOE to actually highlight that point. And the last thing is, don't confuse enthusiasm with capability, because then you open yourself up to attack. So, I'm going stop there. Hopefully, it wasn't a complete random stream of consciousness, but I'm just telling you, this is a very tough environment and we're all going to have to work together to plot that way ahead to success. So, over to you guys for any questions or comments. I'll tell you, you can push back; I take dissent—I don't take it very well. You can ask Colonel Grant, so I'll just warn you.

Q&A Segment

Larrymore: Sir, I'm a cadet, Andrew Larrymore from Texas A&M University. I'm the commander of the cyber operations special unit at Texas A&M University, and I'd love to discuss more with you about what we're doing and perhaps afterwards. But to put it briefly, I'm in charge of ensuring that all the cadets under my command are properly trained in cyber to ensure that as soon as they enter into the military, or they enter into the private sector, that they're able to hit the ground running. They already know a lot of what the cyber landscape is about; they have all their basics. And, so, they're able to immediately start specializing in what they need to do.

And I wanted to ask you, from your perspective, what is cyber lacking nowadays that we can start instilling in cadets and young people now, so that as they grow up in cyber, they're able to start with that base and continue to develop on it?

Maj. Gen. Fogarty: The perfect cyber operator is a language analyst with technical capabilities; that's like the gold standard. Now, the reality is, we have very few such. There are some, but we have very few. So, that's why we had to build teams. In some cases, it may be a little slower, but maybe I don't need the cyber operator to have the language; I can pair them with the language. This idea of being able to contribute as part of a team. . . Leadership, that's the basic issue; that's how we're going to be successful, because they are like unicorns, what I described. Dr. Wetherington probably would agree that the perfect cyber operator is a Language Analyst with the technical credentials.

So, this is where leaders are going have to bring multiple tries, multiple disciplines, multiple capabilities together to accomplish the mission. That's why I'm so excited actually by what you're doing here, because you've really captured the big idea, I think. There was an article that was written a couple of years ago that talked about a Cyber tab versus a Ranger tab. And we had these very inane discussions about, "Well, does that mean we put them in, like, a building with a leaky roof, then we beat on the tin roof, and we keep these guys up?"

No. It literally is about the big idea, Ranger School; which is to do what? It is to create leadership, put you in demanding situations, give you impossible tasks, impossible timelines, and then work as a team. One of the things that we talked to our instructors about, when we ran our cyber branch officer basic course, at Fort Gordon; we brought them all in, and I said, "One of your biggest challenges, when we looked at the resumes of the students that we brought into the branch, is it's mostly these guys are smarter than you are. You have to have the humility to actually recognize that and exploit that."

Many of our cyber operators are smarter than some old dudes like me. What I've got to be able to do is not try to keep him or her down; I've got to be able to leverage their enthusiasm, their capabilities, make sure they have what they require—that's leadership. And then we need to develop that within the technical piece. The good news is we have really great people that are coming into this. What we've got to do is give them the training, give the opportunity; a lot of it has to be leadership. I think that's where we probably are not . . .

We have that out of balance right now, a lot of this keyboard work. Some of us in analysis, critical thinking, we need to invest more in that, and then in the leadership aspect. Because what I want is just like on the Ranger School squad, because you have all different ranks, different services, different specialties. Your ability to be successful in that environment is your ability to effectively lead and recognize the strengths and capabilities of everybody else in that squad.

Because you don't know everything they know, but if you can effectively lead, then you can harness all of that knowledge, all the capability to create the desired effect.

Brooks: McKinley Brooks, I'm a student here at UNG. And in the previous panel discussion, it was mentioned that Russia or China, their interests may not be served by bringing down the United States energy grid in most cases, but it may be served by attacking a country like Ukraine or the Philippines. Do you foresee a way in which the United States could extend a cyber deterrent over our allies in the way that we've extended a nuclear deterrent over our non-nuclear allies?

Maj. Gen. Fogarty: So, that's a really good question. I think, actually, that's the big idea, and it's so expensive to develop full spectrum capabilities and, frankly, we can't do it by ourselves. I can think of operations we're running today; I have high dependency on partners for, on commercial and foreign partners. So, what I've learned at least since I've been in this business from the intel side now to the cyber side is that, this is the key, it becomes the coalitions.

Some cases the coalition is willing; sometimes it's much more formal, and this doesn't mean that you're going to give up your sovereignty. In some cases, they have authorities capabilities that we don't have. If we're working against the common objectives, it's kind of like a Ranger School analogy. Let's supply the authorities and the capabilities that are most relevant to that problem set, regardless of who owns them.

We're going to do it ethically, we're going to do it with our legal framework. But the reality is, that is the secret. And for those that are, they're really trying to get their hands around this, this is one of the things that they're going to have to consider, some politically—that becomes almost an overwhelming challenge. A lot of times, the defense is pretty easy to agree on; it's not easy to implement, but it's pretty easy to agree everyone wants to have some capability. It's the offense, is where people struggle, and then there's a certain degree of defense. If you don't have all the intel capabilities some of the big players have, then that becomes a challenge.

Now, scale is not as important as quality. And what we find is, frequently we have partners that have very unique capabilities, very unique accesses and insights, that are of tremendous value to us. And so, it is all about partnership, and we're doing it every single day.

Lt. Tyme: Good afternoon sir. Lieutenant Tyme, instructor at the Citadel. . .. I know the Navy right now is pushing for all their all-star quarter to have a minimum of two years of instruction in Cyber Warfare and Cyber Security. I haven't heard of much from the other services, specifically; I know some colleges have talked about making it a GeoNet course for degree completion. Do you see value in that, or do you see other schools in industries and services move in that model?

Maj. Gen. Fogarty: Yeah. I think we have to be careful; it's kind of one size fits all. Every soldier in the army doesn't need to be ranger qualified. There are certain techniques, there are certain capabilities, that are useful, I think universally, so that becomes the challenge. What is it that everybody needs to know? And then, what is it that you need to start making the circles a little bit smaller, need to know; if not, you could find yourself here where everybody's in school. Nobody is on the ship, or nobody's in the airplane or are out in the field.

We've corrected this, I think, in the army pretty hard. And that was part of the discussion we had. What do you need to know at the basic course? Basically, what does every soldier and army civilian need to know? That's where we start at, and those things were pretty basic. Then we kind of went up to the upper piece to this. We're just starting; we'll be very interested to see if that's the approach the Navy is going to take; we'll see how that plays out for them.

Neal: Thank you, Major General. My name is Neal, and I'm from Canada, and I have a question. Canada is pretty vast and maybe best from Ottawa, Canada. You're . . .

Maj. Gen. Fogarty: I was in Kingston, two weeks ago.

Neal: I have a question regarding your ethically and legally tackling the challenge of our adversaries that are coming forward. How do you feel, and what are your thoughts? Probably because that's all I'm really speaking to is your thoughts, on maybe your adversaries not behaving the same way or on the same level playing field.

Maj. Gen. Fogarty: So, I'll give an example. One of the reasons they were, I think, were so fast and the Ukraine is, they're not spending three or four days over the target building pattern of life and trying to . . . They are using different rules of engagement. And so, I'm going to tap and key onto the target; I'm going to get good enough. Now, maybe I don't even put direct observation; maybe it's a social media post, and I look at that, and I go, "Yeah, he just said this on Facebook, sent the note to his mom, shows him sitting on the hull of his GMP, there's a geo code, and there's a date time stamp." That's close enough.

I don't have friendly-force scenario; I'm going to shoot. Because it's not precise, it's not kind of what we would consider high confidence on our side. What he does, instead of using a precision munition, he uses a battalion fire of Multiple Rocket Launchers (MRLs). He's putting a couple of thousand bomblets over about a kilometer square area. So, a couple of things happen. One is, he's probably going to get his target, because he's moving faster with adequate accuracy; there's tremendous shock effect. By the way, they've done that. It really sends a really strong message.

Now, the fact that maybe there were civilians in the area, that maybe there are observers from international organizations in the area, they're probably not as constrained by that. I don't think that's the model that we want to adopt. It's what sets us apart, I would argue, from the competition. We're not perfect, we do make mistakes, but I'll tell you, we work really, really hard from my past life in minimizing that. So, no one is perfect, we don't claim we are; we try to get as perfect as we can.

We don't see that same level of effort or the hand wringing after the fact from some of these. That's not the model we want to adopt. Because it isn't long term. That absolutely works against what we're trying to achieve, of who we are or who we aspire to be. But what we've got to be able to figure out is, how to get faster, how to exploit that frankly against them. So, if they do that and you can't let them have a free ride, and you got to understand that's the information piece of this.

When we make a mistake—because trust me, we make a mistake—they are all over it. They amplify; they magnify it. I worked for a guy for a few years whose motto was, "Be first with the truth." Matter of fact, as an intel, as his J2, that was my task, be first with the truth. There is this constant tension of reporting, kind of man bites dog or being able to say, "Hey, we're seeing indications of this right now; I've got low confidence, so I can't confirm, but I want to give you a heads up: this is developing." I think they're on the wrong side of that curve, I think it hurts them.

Now, if they are uncontested in the information space, what they're able to do is suppress all that; they're able to lie about it, and they're masters at that. We have to contest them in that space, and, if we don't, then they get an absolute free ride; they get 100% advantage from an unethical way of operating. We can't allow that, I would argue.

Moderator: We'll take one more question.

Audience Member: Thank you very much. I've very much enjoyed your presentation. You mentioned the need for effective and temporal responses. How often are we really confronted with a problem of clearly identifying the target, trying to avoid unintended consequences and not wanting to reveal our own capabilities?

Maj. Gen. Fogarty: So, that is a daily conversation as I'm up here right now, where I guarantee you, there are some Operations Planning Teams (OPTs) sitting up at Fort Meade or sitting at Fort Gordon, Georgia, that's wrestling with that. We may have a technical capability; we have inter-agency partners and foreign partners who have equities. So, we're going to de-conflict with them. We're going to operate in support of a commander. So, this is not CyberCom doing cyber for itself; this is in direct support of that command.

And frankly, I think you would not only be amazed, but you would actually be gratified by the amount of effort, and that's part of the culture. That's what I will tell you is, when the Snowden disclosures popped out and everyone wanted to jump on NSA, what I think enabled the agency to really continue to operate today, the way it did before Snowden, was the fact that, it's not a lawless culture. If there was a mistake because mistakes will be made, people will fat finger in the wrong IP range.

For a variety of reasons, something won't happen, and what they're trained to do is recognize the mistakes, report it immediately and then mitigate. And there's a permanent record to that. Really, the only way you get in trouble is, if you try to cover up something or lie about it. Is that fair, Dr. Wetherington? We'll retrain you if you made a mistake, so you don't do it again. But if you make too many mistakes, then maybe you can get to something that requires maybe a little less precision or less stress. But the reality is, that is our culture.

And so, when you get the young officers together, many will have a lot of operational experience in the field; a lot of our civilians do also. So, we are not sitting back at some ivory tower talking directly to the supporting commanders. We do it for kinetic operations; we take the same care in cyberspace. That's one of the things, as I look back over my career, it's really satisfying to see that that's who we are at our absolute core.

Moderator: I think we can appreciate them; the problem I have is, it keeps telling at the back of their minds, we have capabilities that we don't use, for reasons of not wanting to reveal those capabilities to adversaries, and that gets in the way of the problem that you identified in terms of effective temporal as well.

Maj. Gen. Fogarty: So, here is what I will tell you, we had a tendency for a while to take our most sophisticated capabilities and employ them as the problem of the minute. And you only have so many of those big capabilities. So, what we found is, we were expending $10 million rounds on a $10 matter, so that's part of it. We want to arrest that. In some cases, the access is so important to us, we are going to deliberately make a decision to actually let it pass by us. And we do it kinetically also.

There are bad guys I've watched for days and we've had an opportunity, and you get a malfunction on the platform. You have a cloud that comes over, so you start to lose situational awareness. Somebody just drives into the X, and we don't pull the trigger. And then we re-cock. We may not get another shot for a year, six months, maybe hours, but that's where the disciplined militaries are . . . It's really taking a long-range view.

I want to be able to do as much as I can, but that's why it's not a CYA; it's literally, I believe that sometimes you have to have that patience, because you can unhinge the whole thing by making a bad decision, when you actually know the facts. So that's the challenge, that's what we deal with every day. Otherwise this will be really easy business.

6

Leo Scanlon

As presented at the 2017 Civil-Military Symposium
Hosted by the Institute for Leadership and Strategic Studies
University of North Georgia

Abstract

The cybersecurity crisis in the healthcare sector is acute, and the recent WannaCry and Petya/NotPetya incidents show that this is a matter of national security concern. The heathcare sector has unique problems and challenges that have shaped its threat landscape, and made it the fastest growing target for cyber attack in the last two years. The cybersecurity challenge for the sector is to overcome the legacy of mere regulatory compliance and embrace dynamic risk management practices as the basis for cybersecurity planning. Critical to that process are the emerging public-private partnerships for threat and indicator sharing, and the legislation that encourages private sector entitites to collaborate with government agencies in making threat indicator information and analytics available across the sector.

Remarks

One of the most fundamental challenges that I think the country faces today is managing healthcare information, and I want to talk to you about where that is going and highlight some of the problems that we're facing, and give you an up to the minute "from the front lines" view of something that General Fogerty mentioned this morning, the WannaCry event, and what that really did to the healthcare community both in the civil public sector and in the government. But first, I'll give the obligatory disclaimer. I am speaking here as a private citizen. My views are my own and not the views of my agency, or the US government.

I think most everybody in the room really understands the big picture genesis of these problems in cyber security, but I want to reference them because they're very important to keep in mind when we start to look at the peculiar features of the healthcare ecosphere. There is a unique quality of the cyber economy. We have gone, in less than a generation, to a point where standalone electro-mechanical devices—computing machines—have became integrated into a largely unregulated public utility which suffuses every single aspect of the economic, military, intelligence and healthcare delivery apparatus in the society today. This went at such a speed that it shocked even the visionaries who put this whole business in motion—and we have not caught up with these changes yet.

The implications of this transformation triggered several major policy decisions that were taken up during the 1990's. The first one was based on the recognition that we had to exclude the software development world from the plaintiff's bar—the only thing that we, in our economy, use to assure product quality. And whether you like it or not, or whether you think it's the best way to go or not, that group of sharks is what keeps everybody in line, whether it's an automobile manufacturer, your breakfast cereal producer or almost any other item you can buy or use, there is a way to adjudicate poor quality. We exempted software development from that mechanism because we understood, based on watching the speed at which software development was happening, that to put software in that particular box was going to undermine the dynamic that was driving this new technology. People were looking at this technology and saying, "We're going to let it happen, and we're going to let it rip," and we did.

Second thing that came along is policy guidance that came as a result of studies done in the military intelligence community, where there was an attempt to answer the question whether we could build a secure infrastructure at the coding and transport layer. The conclusion was that the speed of development in the commercial software world would overwhelm such an effort. It simply takes too long to validate code and products at that level, and the demand for capabilities developed in the commercial "insecure" world would drive adoption of those products no matter what.

Of course, we have JWICS, we have SIPRNet, and we have secure enclaves. But fundamentally, they run commercial products, with some notable exceptions. It was recognized that there would be no way, that except for those highly specialized purposes, could we create truly secure code and software from the base up, nor could we maintain it in a way that would keep up with the activities that were shaping the economy and the use of those technologies.

There was an important series of hearings held by the Congress in the late 1990s and into 2000, where these issues were taken up. On May 2, 2002, Congress held a hearing on the Federal Information Security Reform Act, the predecessor of the Federal Information Security Management Act (FISMA) Dan Wolf, the Information Assurance Director of the NSA testified and outlined the framework for managing the risk of these new technologies in two short paragraphs:

> We suggest that the committee consider assigning a high priority to the development of a comprehensive standard

> for federal system risk assessment and management. The standard should describe—not only the assessment process and documentation requirements—but also include standard methods for characterizing adversarial threats and capabilities, determining categories for mission impact and offer a method for ensuring that the assumptions used in the risk assessment are adjusted as appropriate over time.
>
> A risk assessment—in an interconnected world—cannot be simply completed at the time a system is certified and then filed away. It must become a living document, a sort of trusted calling card that is used when two systems are negotiating their interconnection. The quality of the risk assumptions, calculations and decision thresholds cannot be safely left to chance or independent decisions. There must also be a common method throughout the federal government for managing system interconnection based on a standardized approach to risk assessment. Otherwise, the weakest link in the chain will most certainly break.

These paragraphs defined a roadmap for what would become a decade of effort to develop what we now know as the NIST Risk Management Framework (RMF) and Cyber Security Framework (CSF). The immediate outcome of those hearings was a bill called the Federal Information Security Management Act (FISMA), which instructed the executive branch of the federal government, to use commercial software, use outsourcing for services wherever it is feasible to do so, use commercial network capabilities, and manage the risk that this entails. And the Congress told the commerce department through the National Institute of Standards and Technologies, to develop a series of implementation guidances that will help people understand how to manage that risk.

The Secretary of Commerce has the authority to declare NIST guidance (and other standards used in computing systems) as a

Federal Information Processing Standard (FIPS), which means that use of the standard is a legal requirement for computing systems used by non-military government agencies. The NIST guidance for securing computer systems was called the 800 series, and the primary instructions for and list of controls (800-53) was designated a FIPS, and cybersecurity became a legal requirement in the federal government.

From that moment on, the fundamental challenge in cyber security has been to move to the risk management approach, and away from a static "guns, gates, and guards" approach. As Mr. Wolf noted, the strict compliance regimen that can be put around a standalone environment doesn't work for a networked environment.

The early stages of making that transition were very difficult. In the commercial sector, we started trying to do this when there was no guidance, before the NIST 800 series, before FISMA. There was nothing but the DoD Rainbow Series. These were the guidances that were developed by the military that were structured, controlled, very strict, very robust, but they simply didn't apply in a commercial environment, where you had basically a bunch of young hackers developing web tools to get out to market tomorrow, and the next day, and the day after that. Try lecturing those kids about structured security, and access controls, system authorization and all of that, and all you get is blank stares!

So we had to think much more creatively about what risk management meant, and how to put programmatic structures and controls around this process so that security would not interfere with the business mission. The real thing we had to know was what is the state of our risk—at this moment—and what mitigations did we need to manage it. More importantly, we needed a way to calibrate our tolerance for risk, because we are in a world where we are constantly tolerating and accepting risk. If we don't know what our tolerance is, and we don't know what the threat landscape is, we can get into trouble. But if we can know those two things, we have a hope of being able to manage these technologies in a responsible way. That has been the story of what the federal government, on the civilian side, and the military side, has been doing over the past twelve or fifteen years since this effort began, and in my opinion, to relatively great success.

I'm very proud of what we've done in the government. I'll just give a plug for one of my heroes, and that is Dr. Ron Ross, who directed the FISMA Implementation Project at NIST. If you've never met Ron, you should take the opportunity to do it. If you're not aware of his work, please take the time to become familiar with it. The elaboration of the Risk Management Framework is a milestone in the history of cybersecurity.

Now let's look at how this is playing out in the healthcare sector. In the aftermath of 9/11, Congress directed the government to develop emergency response plans for the critical infrastructure of the nation's economy. The economy was divided into sixteen distinct sectors, one of which is the Healthcare Public Health Sector (HPH). Each sector participates in emergency management planning that coordinates the actions of DHS and other agencies during the response to an emergency. Cybersecurity is a nominal component of that planning, but in most sectors, it is in a very immature state.

This is particularly acute in the healthcare sector, which has a unique attribute: it is the only sector of the US economy that the federal government had to pay to introduce IT technology. There's an interesting reason for this. From an economic standpoint there's very little value to putting patient data in the healthcare ecosystem onto IT systems. And that's not just because of doctors who don't like to write, also don't like to type. That is part of it, if you ever talk to the doctors. But it's much more based on the fact that patient data has a very peculiar relationship to the economic drivers that propel the digital economy at large—it is very difficult to monetize patient information in the way that you can with other commercially generated data. Almost everything we've done to build out the internet economy is built on the idea that at some point that infrastructure is going to host information that can be monetized—sold or used for a commercial purpose that will justify the investment needed to develop new products and capabilities. Typically, that begins with your email address and it ends with your identity attributes, you exchange them for services, the provider re-sells them to others, and that's how the internet works.

You can't do that with healthcare data. We protect the privacy of that data with laws like the Healthcare Information Portability and Accountability Act (HIPAA), which carries strong penalties for the unauthorized release of personally identifiable information (PII). So, in a certain sense, the only people who can easily monetize healthcare PII are the bad guys

The scope and breadth of the target space this kind of information presents is fascinating. For example, have you ever thought about what kind of cybersecurity is in place at a mortuary? Why would anybody attack a mortuary? Well, guess what's in a mortuary? Your full life history from the day that you are born till the day that you died. Why you died, how you died, where you died; and all that information is tremendously valuable to people who want to create an identity that will facilitate insurance fraud, Medicare fraud, and all manner of related identity theft—even after you die.

I'll give you another one. A hospital system was attacked. Forensics figured out that there was no attack on the financial systems of the hospital. No attack on the PII systems and their databases. The only thing that was exfiltrated were xrays. Everybody's scratching their heads. What do they want with xrays? Well I'm sure you've seen the pictures in the Emirates of the huge skyscrapers and all amazing things that are being built in that area. That stuff is built by laborers that come from southwest and southeast Asia—Malaysia, India, China, and similar places. In order to work in those Gulf countries as a foreign laborer, you have to present a clean chest xray. If you have tuberculosis, and don't have a clean chest xray, you buy one. So even chest xray images is monetizable by someone, somewhere.

But you can't monetize that information legitimately, and therefore. there was limited external, commercial incentive to push IT technology and related software and internet capabilities into the realm of managing patient data. The value of healthcare data is "locked" inside the entity that creates it. There are IT systems that use the data, they just don't share it. This means the rate of development of applications is slow, and there is little incentive to create systems that share data.

So in order to incentivize the adoption of these technologies, Congress passed a bill that subsidized the costs. It was called the Hitech Act. It did a good thing with an unintended consequence, one that led to a larger problem.

The good thing with the unintended consequences was that it focused attention on the fact that there are such things as electronic health records systems and if we use them to manage access to patient records, we can cut costs and we can save lives. This is obvious when you think of it in the context of military medicine. If you can get the record of the soldier who's just been injured on the battlefield to the hospital where that soldier's going to end up, and the medics on the evac helicopter and transport airplane know the medical history of the soldier, the drugs the soldier is allergic to, and other information that is vital to starting effective, targeted emergency treatment...you are saving lives. You can't do that with a manila folder with a bunch of post-it notes from the doctors and nurses who've been treating this soldier before they deployed.

Ditto for the civilian population. So the Congress ordered the HPH sector to adopt these systems and the Center for Medicare, Medicaid Services (CMS), which is a part of Health and Human Services, subsidized the adoption of these systems. This is going on down to the small practitioner, the two-doctor office. It's been going on at the hospitals, and the VA, and so on for a very long time. So that is theoretically a good thing.

The bad thing that Hitech did is that it attempted to ratchet up the punishment for the release of data in a very unfortunate way. This was 2009. It was the time when the VA laptop with 25 million veterans' addresses went missing, and Congress was just lit up over this. And when a problem is that big, Congress will pass a law. And when they pass a law like this, they want to make it stick, and the way they do that is to layer on punishments and reporting requirements. Under HIPAA you could be fined for exposing PII, but if you could show that you "could not have known" that the breach was likely, your fines and audit findings would be mitigated. High Tech stripped this defense from the statute, and put the due diligence bar very high.

This created an incentive for organizations to be very reluctant to discuss vulnerabilities or share information about attacks. It also optimized the value of focusing on compliance and fine avoidance as the center of cybersecurity efforts.These two principles are what dominate the healthcare sector concept of cyber security to this day. The real unintended consequence of this showed up once medical devices began being attached to the networks that supported the records management systems that Hitech was encouraging entities to adopt.

To understand what this means, it helps to know that there are two different types of operating systems in the healthcare sector. One type is financial operating systems, these are your backend office systems that do your billing, do your data management, run your office, and pass information to the insurers or CMS.The other category of systems is clinical operating systems, and they are what we typically call medical devices. MRI machines, CT scanning machines, xray machines are all in that category.

These devices are very complicated. They take a long time to design and develop. They take a long time to get approved, whether it's by the FDA or anybody else, and they are big capital investments with long life cycles. They are intended to last for ten to twenty years. So along comes the Hitech Act, and suddenly, the devices that are doing the calibrations of your xray dose, the anesthesia pump, your heart monitor—everything that's going on in the operating room—all of this stuff is now networked, potentially exposed to the internet, and running outmoded Windows operating systems.

These medical devices had been designed ten years earlier with the assumption that they were standalone devices and always would be. They might be running a Linux operating system on the backend, the part that's actually doing the clinical work, but the user interface that calibrates the activity and controls the device is running a Windows operating system. And those Windows systems are not designed to be patched. Not that they can't be, but they are not likely to be, because they were not designed to manage the risk inherent in being networked. The manufacturers had a protocol which was perfectly acceptable for a standalone machine, which is that every six or eight months the vendor's technician would show up, give you a roll up patch, maybe, if it was needed, and bring your operating system up to speed or something close to speed. But now these systems are online, and they're hooked up. They're vulnerable to threats that evolve by the minute, not by the year.

A second point about clinical operating systems is that they're not considered "IT" in a hospital environment. They're not purchased by the IT department, they're typically not under the purview of the CIO, and they're not part of the security program, if there is a security program.

The lack of security in provider entities is also related to the economics of the sector. In the healthcare economy you've got payers and providers. The payers—the insurance companies, the pharmaceutical companies—have money, and they have security programs that manage risk. The providers get their money from the payers, and the providers know what they can bill for a syringe, a bandaid, or an xray. They don't know what they can bill for cyber security. And so they can't build it into their price structure. They can't build it into their business model.

This is a basic business question that every security manager has to face: "what's the ROI of security?" It is tough for a commercial entity or a government agency to answer, but think about what that means for an entity that's running at a net negative level of profit, and go and tell that group of doctors that they've got to shell out money that they might not be compensated for, and they've got to decide what new medical device they are not going to buy or what patient service are they not going to provide, because they've got to do this abstraction called cyber security, and you're starting to get a sense of where the healthcare sector is today. This is the picture that we're dealing with. It's very, very dangerous.

For a long time, people thought this was a minor problem—who would want to hack into a medical device, except scenario writers in Hollywood? But then we had an event that happened last May, and that event was WannaCry. Now what was WannaCry?

You can get into what the intelligence picture is and I think there's still a lot of debate about who launched it and why. But fundamentally, it was an attempt to re-use a tool that had been stolen from the NSA and released to the public and thereby made available to hackers. Someone was experimenting with a way to use it for malicious purposes, and the experiment got loose. The version that got loose was not ready for prime time. It was a ransomware attack, but the blackmail process was not fully functional. Nonetheless it had the capacity to do a tremendous amount of damage to any Windows system that was not up to date. Fortunately, this was not a zero day attack—there was a patch available and if you'd been patching and if you'd been doing anything reasonable, you were ready to handle it. Unless you had an exposed medical device and did not have the access rights to patch it.

WannaCry came on the radar screen in the US early on a Friday morning and was detected at HHS by a newly established threat analysis center called Healthcare Cybersecurity Communications and Integration Center (HCCIC). HCCIC was the cybersecurity interface with the HHS emergency response apparatus that is managed by the Assistant Secretary for Preparedness and Response (ASPR). This disaster response capability is very robust and goes live about seventeen times a year in response to hurricanes, tornados, and terrorist events. There is a similar watch floor maintained 24/7 at CDC and it coordinates emergency response to global pandemics and diseases. HHS has a very robust emergency response capability that is deployed and exercised regularly. But there had never been a cybersecurity component embedded in that apparatus. That is the gap the HCCIC was created to fill.

The HCCIC watchfloor was connected to the National Health Information sharing Analysis Center (NH-ISAC), which is the non-profit entity that connects the sector specific agency (HHS) to the private sector. NH-ISAC partners include the largest pharmaceutical and insurance companies in the world, and they have threat intelligence gathering capabilities that span the globe. HCCIC made it possible for this vast intelligence network to inform the Secretary of the damage WannaCry was doing in Asia and Europe and predict the impact it would have on clinical operating systems in the United States.

The Secretary (Tom Price) was a practitioner himself, and immediately understood the threat this represented. He went to the White House and informed the NSC that he was mobilizing HHS emergency response capabilities and alerting the sector to the emergency. HHS was the only agency that did this. It was the first time that HHS deployed its emergency capabilities in response to a cyber threat, and it is probably the first time that any emergency response apparatus was deployed in this fashion.

The first thing ASPR does in an emergency is to reach out and set up a conference call with large organizations that represent different components of the sector. The first Friday afternoon calls had 1800 lines open. Probably 3000 people on the phone, asking for information. On Saturday we had 3200 lines open, with an estimated 10,000 listeners participating. The information they wanted was very simple. They needed to sort through the internet and media noise, and differentiate bad information from good information—and there was plenty of bad information flying around. People were told at one point that the WannaCry infection vector was email. Well, the NH-ISAC already surveyed and polled everybody they could reach and asked a simple question: "does anybody have an email sample?" Globally around the world, the answer was "no." Next question: "Well does anybody have anything to indicate what the vector was? Several labs around the world had proven that the virus was exploiting vulnerabilities in older versions of the SMB messaging service in Windows. So we were able to tell everybody what the threat was, and how to put compensating controls in place if you couldn't update your systems.

Bad information can be much worse than no information in a situation like this, and the business impact of doing the wrong thing can be worse than the impact of doing nothing. This risk is magnified by the "fog of war" that is inherent in an emergency. There is a commercial incentive for some organizations to jump the gun and try to be first with the headline—whether the information they purvey is accurate or not. The HCCIC-NH-ISAC partnership cut through that, because it was a public-private partnership, with no commercial interest other than providing the best and timeliest information available. For example, one large hospital, a regional system, 4,000 beds, said, "We're about to turn off our email because we're going to protect ourselves from WannaCry." We said, "No. Don't do that. You're going to cut youself off from the updates and the information you need to figure out what this is all about." Reliable threat analytics are a critical capability that the government has to provide to this part of the sector.

The small and regional organizations we were serving could not consume automated indicators coming out of the DHS Automated Information System (AIS) system or any of the commercial incident detection systems. They needed simple, basic instructions. "Do this. Don't do that. We know this. We don't think we know that." This is what they needed, and couldn't get from any other source.

The wake up call was that a cyber event is a kinetic event, it is bigger and more threatening than a natural disaster, because even a hurricane is geographically bounded in its impact. This was an astounding revelation for a lot of people. It was a seminal moment at the hospital and other practitioner level. We now know that during that event, clinical operating systems were shut down across the country. And (as Theresa was saying), this was not reported. At the HCCIC we got anecdotal evidence of the scope of this, but we could not prove it because the hospitals don't report these events to the government. They might report a cyber incident if they think that it's a ransomware-related event and they're really getting hammered and they're going to get blackmailed—then they will call the FBI, and maybe DHS.

This is the state of information and threat sharing information in the HPH sector right now. It is less than rudimentary, and fundamentally dangerous—and this was a virus that was easy to control, easy to get rid of, easy to manage. The full scope of this around the country, I couldn't tell you. We don't think we have a reliable quantitative estimate, but we can tell you that it was widespread and it continued for weeks. We heard from organizations that thought they had patched their systems, but when they rebooted a machine that had been missed, or not properly patched, the virus would start attempting to spread again—this went on for weeks as a low-level event.

Fast forward a couple of weeks after that to Petya, NotPetya. This was a serious menace. And unlike WannaCry, it was more narrowly targeted as a nation-state attack on the Ukrainian economy. Companies that were doing business in Ukraine had to use the fiscal reporting software that was targeted; several major companies were impacted heavily—others were spared. The House Energy and Commerce Committee held a hearing where they invited Maersk and Merck to testify about the damage that was reported in their SEC filings. Full scope of that information is proprietary but I'm going to tell you that those companies got clobbered. Maersk was unable to move ships out of many US berths and other berths around the world for several days. Many of our ports were impacted as a result. Merck is still recovering from about 90,000 machines that were bricked, unrecoverable, unreplaceable, in a matter of minutes with a virus that was propagating with a speed that exceeded the OODA loops of the sandboxing software that they were running on their endpoints.

So this is a serious, serious threat. If a virus like that had gotten loose and had propagated itself throughout the clinical operating systems of the healthcare economy in the United States, you begin to have a threat scenario that until now, nobody has really thought through. And that scenario goes like this. All of the existing emergency management planning for healthcare emergencies is premised on a kinetic event. A tornado is geographically bounded. An earthquake is geographically bounded. You have the ability to move patients from place to place to draw resources from one place to another. You can't, though, move a patient from an ICU into the local Elks hall or the high school gym the way you can if there's been a flood. And if regional or extra-regional capabilities are gone, you do not have an ability to move anybody to get to an operating system that can provide them service.

And this gets to the third element of the challenge that's out there. You don't have to shut everything down to create a catastrophe of this size, even though a catastrophe of that size certainly obviates all extant triage planning. To do real damage to the country, you don't have to do that. You only have to do enough to undermine the public confidence in the healthcare delivery system. That is a major national security issue, and it's on the agenda. It's in front of us today. This is the threat that we're facing at the nation state actor level, and at the hacker level. These things are possible.

These things are possible to do by mistake. It has been recently reported that 911 systems in a dozen states were closed down by somebody who was fooling around, hacking an iPhone. A university student who knew just enough to be dangerous. His program ran a denial of service attack that shut down access to 911 in a multi-state area. So these are very real, very kinetic effects that are coming from cyber security challenges that we are only beginning to learn how to manage.

Some of the wide scale effects I have described are still theoretical, thankfully. We have not done the threat modeling that validates the hypothesis that I presented to you about regional shutdown and we believe that this needs to be done. This is a major effort that's going to require a lot of cooperation to conduct realistic simulation exercises. We've been talking to DHS about infrastructure protection scenario modeling, and there's a great interest in pulling this together as a national project.

The numbers tell a story. In 2010 to 2013, there were 949 breaches, with 29 million patient records stolen. From 2015 to 2017, 113 million patient records were stolen in breaches. The problem affects 90% of all hospitals. These individual records sell for about 20 dollars to 300 dollars on the dark web, depending on what's in them. The impact of the "steady state" attacks is climbing past billions of dollars already, just in theft and recovery costs.

The secondary consequence to the hospitals, of course, is that when they get breached, and they report the breach, HHS, which is a regulatory agency, investigates and sometimes fines them. This is a very significant impact to them and it's just part of the business model. There are hospitals paying ransom everyday because they haven't yet built into their infrastructure, the capacity to have the backups that will let them keep their systems online when they get attacked. Secure off-line backup is basic—but it's a cost that has been avoided until now.

It's going to be a large scale capital investment for these organizations to be able to bring themselves up to the speed of even a moderately secure private sector company or government agency. That's not built into their funding model yet, so there's a lot of work that has to be done at the congressional level to get all of this stuff thought through in a way that will continue to push this transformation from a compliance mindset to how do we get ahead of this problem?

We have also spoken to a number of local and regional fusion centers about the need to integrate cybersecurity into emergency response planning. Fusion centers, as you may know, represent a longstanding approach to bring together emergency management and law enforcement people to share intelligence. In several of these fusion centers, most notably Los Angeles, California, New York, Massachusetts, and New Jersey, they have begun to bring in cyber security resources.

Now this is all in a very rudimentary state. We've visited a couple of these and essentially what they're doing at this point, is reaching out to the law enforcement and asking if they could lend resources from their cyber crimes units. Typically the state police, or the local FBI field office has a cyber technician, and that person will be detailed over to the fusion center on some basis. This is a great start. It indicates that people are thinking, but it is absolutely and totally inadequate to plan for and manage an actual event of the type we are anticipating.

The fusion center model is not the only innovative approach that is gaining traction. The military and intelligence communities have been developing cyber threat analytics centers. They're like fusion centers but they're devoted to cyber threat analytics. The Congress is encouraging the civilian agencies, and the private sector, to follow this model. There are real hurdles that have to be overcome for this to work. In the wake of 9/11, we heard a lot about the transition from "need to know to need to share" in the classified world. This is easier to say than do, but it is essential if we are going to match our opponents. In the private sector world, there are real concerns about liability exposure and the protection of trade secrets, as well as reputational losses that can occur if confidential threat information is shared.

That is why, in the CISA Act, the Congress said that private sector entities may share indicators of compromise and threat information with Federal, State, Local, Tribal, and Territorial agencies without fear that that information will be used for regulatory purposes. It's a big step and opens the way for organizations that have been sharing privately, behind the scenes, to do so in a structured way. But here's a problem. It's been estimated that the typical threat analyst talks to and shares with about five other analysts. These highly informed conversations occur among people who are not simply watching signatures, but are really trying to figure out what the anomalous events that they have seen mean in context. This is analytic work—the same attack tool is used by many actors—understanding the intentions of the attacker—context—is critical to containing and eradicating the intrusion, and minimizing damage.

It's the human in the loop that takes the automated information that comes in off of the log servers and firewalls and various other devices and threat feeds that you can buy or get from DHS. That fire hose of information doesn't mean anything out of context, and who creates context, at this point, are human analysts. This trust-based intelligence sharing is fundamental to building an early warning radar about things like WannaCry or Petya or other techniques that are incipient and maybe being tested out and not yet fully deployed.

This is an intelligence capability that can't come only from the government. The military intelligence community is not tasked to dig down into and understand context in the commercial sector at this level. This is a task that the commercial sector has to take up in conjunction with the public sector, and where I think the most important future developments are going to go is public-private partnerships. We're talking today about civil-military collaboration in cyber security. The next layer out is public-private coordination through the ISACs and intermediary institutions like the HCCIC that need to be brought into existence across all sectors of the economy.

There is no way that any one entity or part of this whole structure can defend itself from the types of attack and the ferocity of attacks that we're seeing today. The bad guys share with each other. They crowdsource their solutions. They buy and sell. They have no compunction about it. We should take a lesson from them and do exactly what they're doing and do that on a scale that allows us to get into the game. And that's the only way this is going to work. That's the only way it's going to work in a sector like the healthcare sector, but I believe that it's fundamental for the entirety of the commercial sector of the United States.

This is fundamentally a complex social problem. We've talked today, and other people have pointed out, that at the root it's not a technical problem. There are technical elements to it and the technical elements are very, very complex. We are going to have to figure out ways to get elements of the society that have different interests, sometimes even conflicting interests, to find a path to the common good. We have to square the public interest in agencies being able to exercise regulatory responsibility with the need for regulated entities to share information that can protect others. That's a matter of generating trust, trust in our public institutions, trust in the collaboration between the organizations that carry out this battle. That development of trust is really a matter of leadership because without leadership this will not happen. If we do not think, in our minds, that our job as cyber security leaders is to inculcate, develop, and build this sort of trust in this ecosphere, no technology is going to solve this problem.

The last thing I want to tell you about is something that was done through the NH-ISAC and which I think is a very exciting initiative. It's the role of volunteers—crowdsourcing solutions. Crowdsourcing works, and not just for the bad guys.

You may know about the federal research corporation called MITRE.They're a not-for-profit entity that does difficult thinking about very hard problems that the government doesn't have the experts in-house to do. MITRE studied, and then tested out at the NSA, a detailed mapping of how does an attacker gain entrance to, and persist undetected in, a Windows network. The outcome is called the ATT&CK Framework. It's a "periodic table of the elements" or a map of the digital genome, if you will. But, the map of the genome it doesn't mean anything if you don't have a drug, or a protocol, or some procedure that lets you use that map to address an organic process you want to alter. Just so with the ATT&CK Framework—the task was to develop tools that would use the map to detect intruders.

This was briefed at a summit meeting of the NH-ISAC, and one of the partners said, "Look. I do not have the money to figure out how to use that, but that is valuable, and it's valuable to everybody. What if we, all of us, just start doing the research. We divide it up. Each of us figures out what data we've got. What tools we've got. Start a study, develop queries that can be run against large data sets looking for these small indicators, and we make it a collective effort. We make it public, and we make it free. Everybody gets in. Nobody can profit. The output is free. Everybody benefits." Ten companies signed up that day. A year later there are thirty-three companies devoting a significant portion of their top talent, the folks that they've got who are serious research analysts, malware analysts, they're working through this problem and working on developing solutions.

We were talking about this to colleagues at the Financial Sector Advanced Research Center (FSARC)—which is associated with the Financial Sector ISAC, and it turned out that they were developing a tool to tell you how accurate that query is. Just organically, we in the healthcare sector were coming at this from the bottom up, the finance sector was coming at it from the top down, all with the same objective, and all on a volunteer basis. This is a powerful model, where funding from the corporate sector, with a little bit of support from the government, facilitates a major research project which is designed to produce cybersecurity resiliency to the entire economy.

This is a model that we are seeing work in social media. And as I said, the bad guys have figured it out. I am very optimistic that at least at these levels, core elements of what we need to do are in place, growing, and are going to be understood, and are becoming more and more a part of the daily life in cyber security

Back to the point of leadership. Bruce Schneier, the well known security blogger, has a great quote. It says "amateurs attack machines, professionals attack people." The job in cyber security is about leading people. So I'm really thrilled to be part of this summit and to be here at UNG because it's clear to me that this is the mission that UNG is taking on and succeeding at in a tremendous way. So with that, I want to thank you very much. It's an honor to be with you. And if you've got any questions, I'll be glad to talk as long as they let me.

Q&A Segment

Audience Member 1: Thank you very much for your talk. It's more of an observation than a question. Many of these issues, especially with legacy systems, and the lack of patching, I know sometimes in private industry you can't patch systems because the system itself will fall down. How do you incentivize private industry to come to the government and say we want to share information, and how is that then shared amongst government agencies and internationally, amongst friends and allies?

Scanlon: Right, so, this is really the 64,000 dollar question. I don't think you need to incentivize people to share information. What you have to do is make it clear to them that they will not be punished if they do that. That they will not find themselves in a regulatory trap or otherwise have an adverse impact of doing it. They want to share information and in fact they are. As I said, the analysts across these different companies and they've got their buddies, and they've got deep reach back into the intelligence community—this information flows around. People talk. But they do it very quietly, carefully, and in ways that are sure to keep the risk managers, and the lawyers out of the conversation. That's good, but totally inefficient. So if we can remove the remnants of those regulatory barriers we can grow this model of threat analytic centers and encourage them in various communities of interest. This is what the Information Sharing Analysis Centers (ISAOs) are supposed to do in the US, for example. Government has to point the way as well. Recently the NH-ISAC was awarded an openly-competed grant from HHS for the purpose of supporting a public portal that would be available not just to their members, but to anybody who wants to share threat information or get threat information from them. That is also shared with the government through HHS and with DHS, and DHS has the NCIC.

So what we're developing here is the model . . . the old model that we're kind of moving away from would be a hub-and-spoke model. In that model, some piece of information goes up, and some smart guy at the top says "oh, that means this" and then it goes out to a subset of recipients. What we're moving towards is a mesh. So what we want to enable, and the CISA Act in fact enables, is mesh communication for an important category of cybersecurity threat indicators. Without getting too technical, there are liability issues that the lawyers are still sorting through, but the mesh model has been instantiated in the law.

So the word is out there. Start to do this. The collaboration internationally is straight up. There are threat analytic centers in Europe. And these things are popping up all over the place. Like I said, down to the local and state level in the US, it's being put together. So I think this is beginning to happen and it's happening organically. Now, the question becomes what type of communication and how can we develop context for that. So different sectors have different context models that need to be developed.

Audience Member 2: Thank you. So you mentioned that the threat is real and you describe how it's not just a cyber threat, it actually transcribes into a serious threat to healthcare, and services, and information, and people, and our livelihood. We've discussed in different ways, that this threat is real because we have poorly-architected design, poorly-implemented systems, or poor cyber hygiene. So while we chase the latest technologies, what are your thoughts on how to get industry to improve, possibly through government policy?

Scanlon: That's a real tricky question and there's a lot of room for debate. I'll just take the medical device issue. The manufacturers are begging for a standard that would give them a floor that they can build to, so that they can start introducing cyber security into their design. The government is wary of doing that for a lot of reasons, not the least of which is that very often mandated standards don't work for very long. So the question is how do you induce self-regulated development in the private sector? I would say that the government can help broker that conversation but that's a matter of leadership. That's back to having the will, the vision, and getting people with competing interests to begin to sit down and say "listen either we all survive or we all fail." And that conversation is where I think the solution to that problem will come.

We could force good cyber hygiene. But a zero day is a zero day. And as good as your hygiene is, as compliant as you may be, when somebody finds that line of code in that 200 million or whatever it is lines of code that's running on your desktop, and nobody knows about it—bingo. The bad thing is going to happen. So resilience and response are the most important things we can do in addition to building the collaboration that starts to get the industry itself to police itself. I think there are models for doing that. It would be a longer discussion. There are models for industry self policing that work and I think we could drive that forward as a big part of that solution.

7

Bill Smullen

As presented at the 2017 Civil-Military Symposium
Hosted by the Institute for Leadership and Strategic Studies
University of North Georgia

Abstract

The threats and challenges facing America and indeed the world today are many and they are serious. The list is long and includes cyber-attacks, which are an existential threat and a national-security menace. They can undermine democratic institutions or democratic governments.

There needs to be a clarion call for vigilance and action against the threat of cyber-attacks, which transcends the public and private sectors. Cybersecurity controls are necessary in both and insufficient in either.

If cyber threats are the new normal, we must slow or stop their progress. We need a grand strategy that calls on civil-military cooperation and international collaboration that shares knowledge and increases defense effectiveness.

A grand strategy is not especially esoteric. It is the calculated relationship of determining means to ends while addressing the what and how of responding intelligently so as to advance vital interests of all those threatened by cyber-attacks.

The New Normal

I spend a lot of time thinking about national security. It is what I do professionally. My National Security Studies Program has had its home at the Maxwell School on the campus of Syracuse University since 1996. A premier professional development program, it offers executive education courses for senior civilian and military leaders responsible for the national security interests of their respective organizations or agencies.

My obligation to them is attempting to determine the threats and challenges to the United States and indeed the world at large. It also includes determining appropriate responses.

The challenges facing this country and the community of nations are many, and they are enhanced by our contemporary adversary – "uncertainty." We live in a world without rules or reason. The global order we came to accept during the Cold War period is under immense strain, some say even in partial collapse. Threats to our safety and security as a people are serious. My list of threats is not especially long, but it is weighty.

At the top is terrorism and the sixty terrorist organizations, led by ISIS, listed on the State Department's website. The outgrowth of homegrown terrorism is equally alarming.

With respect to worldwide threats, I have a short-term and a long-term list. Both are fact-based and troublesome.

My short-term list has North Korea in a position of prominence. Kim Jong Un's aim is to perfect a nuclear capability so that he can launch an intercontinental ballistic missile capable of reaching the United States. He says it is not a threat, but a reality. In reality, it is a threat that could lead to potential military action, including nuclear war.

My long-term list is occupied by Russia, which has an aggressive president, Vladimir Putin, seeking to reinvent that Cold War empire we faced for decades. The Soviet Union may have disintegrated in 1991, but Putin has chosen to nibble at pieces of the fifteen republics that were once part of the USSR.

My troublesome list includes China, Syria, and Iran. Each has their own problems, but each has managed to form a troubling relationship with the US. My guess is that picture will not improve anytime soon.

Finally, there is my conundrum list. It has but one occupant, but it is the most difficult to attack, because it cannot be seen and confronted like the others. It is a cyber threat.

Cyberwar is considered a relatively new phenomenon when it actually began in 1967 with the advent of ARPANET, or, as it was more commonly known, the Advanced Research Project Agency Network. It was in that same year that computer pioneer Willis Ware authored a prescient paper, "Security and Privacy in Computer Systems," in which "he envisioned unwelcome visitors penetrating or hacking into them." He assumed "sabotage was possible but espionage was more likely" (Brown 2017).[1]

Today, fifty years later, threats to the real world from cyber are worse than ever, and the situation continues to deteriorate. Cyber-attacks are an existential threat designed to steal consumer information or embarrass business executives and politicians, among other annoying possibilities.

Whether conducted by lone wolves or nation-states, they can compromise the safety of medical, food, and water systems; disrupt transportation systems; or even bring down power grids or destabilize nuclear plants. Such attacks are a national-security menace that can undermine democratic institutions or democratic governments.

There needs to be a clarion call for vigilance and action against the cyber threat. That threat transcends the public and private sectors of society. Cybersecurity controls are necessary in both but insufficient in either.

This information war is currently being waged on some sophisticated territory. It requires critical thinking and people who have the powerful vision necessary to confront the threat.

That requires strategic thinking and strategic planning. They are not one and the same. Strategic thinking looks at where we have been, where we are now, and where we want to be in the future. We can utilize this methodology for cybersecurity. By connecting the dots between the three, it provides a sense of clarity. It leads to strategic planning and, ideally, to the invention of ideas regarding what needs to be done to move the needle of counter-cyber progress.

It is not easy. Creating a cyber-defense and security system is extremely difficult today and may be for years to come. Too many things are changing on the Internet along with devices and connected equipment that move at warp speed. For example, billions of devices already exist around the world and are connected to the Internet. That number will only grow as the population grows and technology advances. Each device is a potential point of attack or a potential weapon in a distributed denial of service attack. As one expert reflected not long ago, "Defending the cyber environment is like trying to change a flat tire on a vehicle going down the road at 70 mph" (Coleman 2017).[2]

While alarmists in academia and politics warn of the threat of a "Digital Pearl Harbor" or "Cybergeddon" potentially paralyzing a connected society, we are not quite there yet (Etzel 2003).[3] Nevertheless, we need to find ways to contribute now to the enhancement and protection of national and international cybersecurity, so we do not ever get there.

That may sound too optimistic, but it is a goal worth having. We need to protect against the likes of the mysterious hacking group that supplied a critical component of the WannaCry "Ransomware" software attack that spread across the globe in mid-May of 2017.[4]

That global cyberattack went by the name "Shadow Brokers" and began freezing more than 300,000 computers in 150 countries.

It wreaked havoc on businesses, universities, and governments. Spread by email, it locked users out of their computers and threatened to destroy data unless a ransom was paid.[5]

Ransomware attacks have increased exponentially in past years, costing businesses billions of dollars. Equifax, a credit-reporting company, experienced a massive data breach in September 2017, compromising the personal information of some 145 million Americans. This personal information included Social Security account numbers, driver's license numbers, and addresses. It is discomforting to think that virtually everyone in the US. could be affected in a data breach in some way, somewhere, over some unknown span of time.[6]

These types of malicious actions further amplify the loss of confidence in information protection. They also create significant problems for companies and institutions responsible to their stakeholders. It is difficult to make decisions when you do not know what data are real and what data have been manipulated.

So, if cyber threats are the new normal in both the public and private sectors, what must we be doing to stem the tide? I am all for a grand strategy that calls on civil-military cooperation and international collaboration, a strategy that shares knowledge and increases defense effectiveness, one with an "all in" mindset.

This grand strategy need not be especially esoteric. It can be as simple as the calculated relationship of determining means to ends, while simultaneously addressing the what and how of responding intelligently so as to advance the vital interests of all who are threatened by cyber-attacks.

Furthermore, the strategy should consider what resources are available, what costs are involved, and the potential consequences, as well as the risks and ramifications associated with the established objectives. This strategy should be operational as well as aspirational. It should not allow a lack of imagination to obstruct its effectiveness. A "can't happen to us" mentality is an unspeakable such obstruction.

I can think of five attributes of a responsible grand strategy:

- First, it must be balanced to ensure certain competing interests—such as security, economics, and values—are properly aligned.
- Second, it must be prudent so as to strike the correct balance between what is needed and what is possible.
- Third, it must be principled and built on a combination of dialogue, cooperation, and transparency.
- Fourth, it must be purposive which is to say it must advance interests and ideals while articulating a positive vision for the future.
- Fifth, it must be sustainable so it can serve as a model that can endure and grow in its effectiveness over time. [7]

Is such a grand strategy failsafe, or will it operate without error? Of course not. But it beats suffering from the strategic thinking deficiency that has plagued us for far too long. If the prospect of a "Cyber 9/11" grows likelier by the day, we need a plan of action now.

Just as there are sixteen intelligence agencies in America, so also there are at least eleven federal agencies that bear significant responsibility for cybersecurity. They include: US Cyber Command, the Central Intelligence Agency (CIA), National Security Agency (NSA), Department of Homeland Security (DHS), the Treasury Department, three branches of the military, the Federal Reserve Board, the Federal Deposit Insurance Corporation, and the Office of the Comptroller of the Currency.[8]

Should they be merged—like DHS was formed in 2002—into a single agency? My dislike for bloated bureaucracies tells me no. But like the post-9/11 formation of the Office of the Director of National Intelligence that forced intelligence agencies to collaborate more effectively, having a Director of Cyber Defense makes sense. He or she could bring the strands of cyber defense together to help create a credible defense of, and a credible retaliation to, cyber-attacks.

This position, if it were to be created, should be led by someone who demonstrates cyber experience and the management skills to bring collaboration of the cyber experts in the various federal agencies to a new level.

Someone who can capture the attention of the private sector, where cybersecurity has not been a particularly high priority until recently.[9]

Someone who can tear down competing objectives that can undermine the country's ability to address cybersecurity challenges, and someone who can work more effectively to weave talent, technological breakthroughs, and early warnings together in a concentrated effort.

Someone who can provide crucial coordination with state and local governments and the business community.

Someone who can enable better international cooperation that can help assist in an improved exchange of information between nations.[10]

Someone who can help build the bridges of coordination and cooperation between the military and civil sectors, the public and private sectors, and the national and international sectors.

Someone who can mine promising technologies, like artificial intelligence, and simultaneously invest in software that automatically detects and thwarts attacks.

Someone who can facilitate the exchange of opinions, doctrines, strategies, structures and best practices.

Someone who could design a digital "immune system" using artificial intelligence to monitor networks for suspicious activity (Eagan 2016).[11]

And perhaps, most importantly, someone who can inspire motion and emotion by heeding the words of Abraham Lincoln who once said, "The best way to predict your future is to create it."

Endnotes

1. John S. Brown, "Cyberwar Isn't So New – It Began in 1967," Army Magazine, September 2017, 65.
2. Kevin Coleman, "Complexity of developing a cyber defense strategy [Commentary]," c4isrnet, June 7, 2017, https://www.c4isrnet.com/home/2017/06/07/complexity-of-developing-a-cyber-defense-strategy-commentary/.
3. Barbara Etzel, "Digital Pearl Harbor," in Webster's New World Finance and Investment Dictionary, ed. Barbara Etzel (Houghton Mifflin Harcourt, 2003), Credo.
4. Chris Graham, "NHS cyber attack: Everything you need to know about 'biggest ransomware' offensive in history," The Telegraph, May 20, 2017, http://www.telegraph.co.uk/news/2017/05/13/nhs-cyber-attack-everything-need-know-biggest-ransomware-offensive/.
5. Olivia Solon, "WannaCry ransomware has links to North Korea, cybersecurity experts say," The Guardian, May 15, 2017, https://www.theguardian.com/technology/2017/may/15/wannacry-ransomware-north-korea-lazarus-group.
6. Lily Hay Newman, "Equifax Officially Has No Excuse," Wired, September 14, 2017, https://www.wired.com/story/equifax-breach-no-excuse/.
7. Coleman, "Complexity."
8. H. Rodgin Cohen and John Evangelakos, "America Isn't Ready for a 'Cyber 9/11'," Wall Street Journal, July 11, 2017, https://www.wsj.com/articles/america-isnt-ready-for-a-cyber-9-11-1499811450.
9. Steven Weber and Betsy Cooper, "Moving slowly, not breaking enough: Trump's cybersecurity accomplishments," Bulletin of the Atomic Scientists 73, no. 6 (October 2017): 388.
10. Cohen and Evangelakos, "America Isn't Ready."

11. Nicole Eagan, "To Stop Hackers, Treat Them Like a Disease," Wall Street Journal, June 15, 2016, https://www.wsj.com/articles/to-stop-hackers-treat-them-like-a-disease-1466000424.

Bibliography

Brown, John S. "Cyberwar Isn't So New - It Began in 1967." *Army Magazine*, September 2017.

Cohen, H. Rodgin, and John Evangelakos. "America Isn't Ready for a 'Cyber 9/11'." *Wall Street Journal*, July 11, 2017. https://www.wsj.com/articles/america-isnt-ready-for-a-cyber-9-11-1499811450.

Coleman, Kevin. "Complexity of developing a cyber defense strategy [commentary]." c4isrnet, june 7, 2017. https://www.c4isrnet.com/home/2017/06/07/complexity-of-developing-a-cyber-defense-strategy-commentary/.

Daniels, Michael James. "Power, Politics, and Process: Issues in US Grand Strategy." *US Army War College Strategy Research Project*. Carlisle, PA, United States: US Army War College, 2014.

Eagan, Nicole. "To Stop Hackers, Treat Them Like a Disease." *Wall Street Journal*, June 15, 2016. https://www.wsj.com/articles/to-stop-hackers-treat-them-like-a-disease-1466000424.

Etzel, Barbara. "*Digital Pearl Harbor*." In *Webster's New World Finance and Investment Dictionary*, edited by Barbara Etzel. Houghton Mifflin Harcourt, 2003. Credo.

Graham, Chris. "NHS cyber attack: Everything you need to know about 'biggest ransomware' offensive in history." *The Telegraph*, May 20, 2017. http://www.telegraph.co.uk/news/2017/05/13/nhs-cyber-attack-everything-need-know-biggest-ransomware-offensive/.

Newman, Lily Hay. "Equifax Officially Has No Excuse," *Wired*, September 14, 2017. https://www.wired.com/story/equifax-breach-no-excuse/.

Solon, Olivia. "WannaCry ransomware has links to North Korea, cybersecurity experts say." *The Guardian*, May 15, 2017. https://www.theguardian.com/technology/2017/may/15/wannacry-ransomware-north-korea-lazarus-group.

Weber, Steven, and Betsy Cooper. "Moving slowly, not breaking enough: Trump's cybersecurity accomplishments." *Bulletin of the Atomic Scientists* 73, no. 6 (October 2017): 388-394.

8

Colonel Jeffrey Collins

As presented at the 2017 Civil-Military Symposium
Hosted by the Institute for Leadership and Strategic Studies
University of North Georgia

Thanks. Thanks for the kind introduction. I'll get the Air Force-Army football game out of the way now. We won't mention it again. Just in case you don't know, the Air Force mission is to fly, fight, and win in airspace and, wait for it, cyberspace. To fly, fight, and win in airspace and cyberspace. Now, the difference in those three domains, one of the differences at least, is that the first two are physics limited. What's the limitation on cyberspace? The answer is imagination. But, not only imagination, but also our ability to imagine a new way of fighting in cyberspace and then to enact that way of fighting in cyberspace.

I'm going to talk today about the cultural change that is required for us to keep up with our adversaries in cyberspace and how at least the Air Force is going after that by standing up CyberWorx at the Air Force Academy. The lawyers back at the Air Force Academy will listen for this carefully on the tape, but these are my views and not the Department of Defense's (DoD) views which are up there on the corner. That's what I'm going to walk you through, and I'm going to talk about it in terms of how we're getting after this cultural problem by delivering capabilities very rapidly back to the Air Force using a project-based learning approach, but also by using gamification and teaching the cadets that it is about their ability to apply imagination rapidly as part of the cyber workforce that's going to determine whether or not we keep ahead of the game.

I'm going to talk a little bit about the problem. We in the DoD are good and have built acquisition systems that are designed to create Death Stars. The title of my boss was Information Dominance, and you can't really say dominance without pronouncing it that way. We have a world in front of us where we need our cyber talent to be able to pick up Legos and build TIE fighters, X-wings—sorry, wrong side of the Empire divide—to build X-wings to take down Death Stars. Our acquisition systems are built to build Death Stars and, yet, we're living in a DevOps world. Not only are we living in a DevOps world, our adversaries are also living in that DevOps world. Your ability to implement imagination and what is needed on a battlefield rapidly will determine your success in flying, fighting, and winning in cyberspace.

I'm going to talk about three problems. One of them is that we have a culture of risk aversion versus a culture of risk appetite. There are reasons to avoid risky behavior—international, political reasons, fiscal reasons, a lot of reasons. However, in the cyber domain, agility brings its own reward and risk aversion versus a risk appetite will get you killed.

We have built our systems to engineer risk to as close to zero as possible, and what that has meant is that we're not as agile as we need to be in this domain. Overcoming that risk aversion where it is okay to fail a bunch of times in order to find out what is going to work, a culture of experimentation and rapid integration of new technologies into war fighting, is what we're after on that first bullet.

Incrementalism versus agility. Somewhere we came up with the notion that it's better to have a little success than to try something bold and to fail and then to try something else. The latter is agility. Being able to take what Silicon Valley is producing and being able to integrate that into your war fighting at Silicon Valley speeds is what our enemies are doing. We, instead, have created budgetary processes and acquisition processes that prevent us, or at least make it hard for us, from doing that integration. We talk about it in terms of a Valley of Death between innovation and integration into war fighting. In the other domains, we can overcome that in some senses. But in cyber, we have to get faster.

Then, finally, the system's focus versus effects. I'm going to drill a little bit into this. Anytime you hear an Air Force officer talk for more than ten minutes, you're going to hear about the OODA loop, which is "observe, orient, decide, and act," and I use as example the fact that you do this when you're driving all the time. You see a squirrel on the side of the road, and you orient yourself. Is that squirrel moving toward the lane of traffic or away from the lane of traffic? You then decide, I guess depending on your character, am I going to speed up, or am I going to slow down and break for the little squirrel? Then you act on that. That happens at a tactical level, but it also happens at strategic and political levels.

Our ability in cyber is focused more on these three dimensions of the information environment that the joint doctrine talks about. Our ability to speed up our own OODA loops and also our ability to slow down an enemy's OODA loops determines whether or not we are successful, not just in the cyber domain, but in all the war fighting domains, as well, given that cyber has, for better or worse, integrated, infused, become a leech, an enabler, or whatever your favorite term is on those other domains.

To take the data dimension as an example, I can formulate a way to get the enemy to not be thinking about launching its next sorties, but instead to be trying to unscramble its targeting data so that it can decide whether or not to launch the next sorties. On the physical dimension, I can break someone's radar so that they don't know where the American Air Force is right now, or I can feed a different picture of where we are right now.

This doesn't have to be cyber. When I was a commander in Afghanistan, when I would be going along with the Provincial

Reconstruction Team (PRT) meeting with the village elders, an F-16 would fly overhead. That F-16 in our proximity was not an accident, and it takes a lot of cyber to get those two things to happen at the same time, a lot of coordination, a lot of making sure that our Supervisory Contral and Data Acquisition (SCADA) systems are ready to refuel that aircraft to get it in the air, to get it navigated to the right place. But what was the impact of that on the cognitive dimension of the information environment? Our enemies knew that they could kill us, but the enemies also knew that the Air Force was there.

Thinking through the impact of information, and you can take our elections last year as another example of where the cyber dimension can impact our cognitive dimension, I think both the Air Force and the Army are thinking through the implications of information warfare, information operations, and I'm going to tell you how we're getting after that, at least in CyberWorx.

I've talked about the problems. Then when I go to this slide, this is the audience participation where you're all supposed to do something like, "Wow. Look at that background. My goodness gracious. Wow." What is the Air Force's answer to how we get over these problems? The answer isn't some big, giant strategic acquisition reform which is a fifty-year term of art. Here's what it is: let's figure out how to change the culture, and let's figure out how to solve problems rapidly, and CyberWorx is the answer to that.

The vision of CyberWorx: you should notice that it's not about technology. It's not about that physical dimension of the information environment. It's about people. What we have accidentally done is we have made it so that our airmen might be able to imagine new things to do but then working through all the bureaucracy to do that thing is too hard and slow. We have not unleashed the power of airmen to create apps rapidly. We have not unleashed the power of the airmen to enact new ways of war fighting as rapidly as we need to. Unleashing that power so that we can really get after cyber, so that we can really unleash the potential cyber power, is what CyberWorx has stood up to do.

I'm going to show you the mission that the Air Force gave us for CyberWorx. That's to accelerate operational advantage. A lot of times in cyber—and these are cyber wings that I'm wearing, Air Force flies, fights, and wins in cyber, so we have wings for cyber just like we have wings for space and wings for pilots. To fly, fight, and win in cyber is about operational advantage. A lot of times people wearing these cyber wings want to geek out on technology. What we need, instead, is a huge leap in cognitive diversity of individuals who are worried about the cyber domain and who realize that cyber is commanders' business. It is not just people wearing these wings' business. It is not the server room technicians' sole business, but all those people have to play a part in cyber.

How we're getting after this to create the simpler, more intuitive capability for the Air Force to enact imagination faster is we're partnering with industries. At CyberWorx, we have created what's called a "partnership intermediary agreement." We can geek out on that during question and answer if you care about how we're actually doing this within the law. But, what it enables us to do is to use a nonprofit in Colorado Springs so that when we get a cyber problem from the Air Force, we do a call through to this nonprofit for industry to come in and help us solve that problem.

Now, the agreement is between the government and the nonprofit and then between the nonprofit and industry, and that's important because intellectual property has to be protected and has to be agreed upon before you start doing anything, and this is where we get to the innovation phase: design thinking. If you're not already familiar with design thinking, then you should spend a little time—not right now, but after I'm done talking or during a break or maybe later—to look up design thinking, and you'll find—a Harvard Business Review article will probably be among the first ones—companies that integrate design thinking into the way that they do business end up being 40 to 60% more profitable than companies that don't, that use waterfall methods or other ways of getting after problems.

The benefit of that partnering and innovating lines is that we then we get rapid solutions. Now, if you want a perfect solution, you can go out and hire a corporation like RAND Corporation or a Federally Funded Research and Development Center (FFRDC), and they will study a problem for a couple of years. They will articulate that problem and give you some recommendations for moving forward. That is not the CyberWorx model. We are not articulating problems. What we're doing instead is we're understanding the problem well enough within a one-week design sprint to give you ways to move forward fast. While we're doing that, we are educating our cadets, because the cadets are in the design studio.

Cadets are included once we decide that a problem is going to be part of a class we teach called the Technology Innovation Management Class at the Air Force Academy. The reason that it's a management class is because we want a cognitively-diverse group of students, so we have eleven different majors in our design studio working on two problems that the Air Force has given us. That notion that bringing cadets into the problem, that the cadets don't know enough about a problem to really come up with solutions, is false. The Gen Z cadets, the millennial, the end of the millennial cadets, they don't bring the blinders that even our captains have learned, and so we bring them into the studio then leaven them with industry partners who have a lot of experience. We've had brand new industry partners, recent graduates of civilian universities, and we've had industry partners who are in their 70s come into the studio to work as a diverse team. You know what? Cadets learn from that. They learn that a way for an officer to get a fast solution to a problem is to bring people together and talk.

In our design studio, we do our best to ignore rank because frankly the three-star general in the design studio does not necessarily have the best idea. It may be the 18-year old two striper. Everybody in the cyber domain is important, and we have to figure out how we listen to them and how we move forward fast.

Then, finally, we build. We deliver demonstrations of the capability, or we deliver a working demonstration. Then we go into a DevOps model. Development operations model is one where you understand that the first release of a product or of a capability is not the last release, but it's just the release that you're going to use today to get after war fighting. Then you're going to tell the developer how it's going to be improved at the next delivery, which is hopefully an hour from now.

This is what our design studio looks like at the Air Force Academy. We talk about education at elevation because you can basically see Georgia from those windows if you look out there. We're right on the edge of the hill. We've got forty cadets in the studio this semester. We've got ten industry partners. Each industry partner is part of one of the cadet teams. We're going through two problems. We also, though, don't just limit our participation in the studio. We send our cadets out.

One of the problems that we were working on this semester is about multi-domain command and control. That multi-domain command and control happens at various points in the Air Force, like Nellis Air Force Base and Langley Air Force Base. We send our cadets and industry partner teams out to do the research, to talk to the war fighters who understand the cognitive dissonances that are taking place that lead us to need a better solution to the particular problem that they're focused on.

We get them out. We then use design thinking. The student who's there in the upper left-hand corner of the slide is doing an outbrief. We create personas of the different airmen that they discover out there that are experiencing the problems, experiencing the pain points. Then we act those out back in the studio so that then the whole team can understand that problem, so then the whole team can design solutions to overcome those problems, whether they're material solutions or just stupid policies that we need to change. You have to be able to articulate that.

What I've been describing so far is really two different approaches. One is project-based learning. You can read there some of the attributes of why project-based learning is a way to change the culture of a whole organization. Now, we're changing it through teaching our cadets, but we're also bringing in operators from across the Air Force into the studio as well so that they also understand how this culture is changing. They take back with them to their units these new methods of solving problems, of getting after capabilities rapidly.

Then the piece I'm going to talk about next is about serious gaming. If you don't know what serious gaming is, it's basically application of simulations and applications of putting people in situations to allow them to learn to be flexible in a place where it is okay to fail very rapidly and very repeatedly. The benefit of design thinking in project-based learning is that you do a bunch of very rapid prototypes with the expectation that most of those prototypes are going to fail so you then know what is going to work. Serious gaming is the same way. You're doing a lot of movements that you know are going to fail, but you are learning from those as you go along. It ends up being reflective and reflexive learning.

What does that look like? The picture that you see in the upper right corner, we call that CyberCity. I know it looks like a train set, and it is a train set. But it also has all the aspects of a city that you would expect, and all of those, for example, power plant and traffic lights and cameras around the city are all controlled by the real control systems that are out of the camera frame in the back.

What are the cadets doing there? They're hacking into those systems, and other cadets there are defending those systems. What their goal is in one of these scenarios is hostage exfiltration. I talk a little bit about multi-domain thinking. If you think about a hostage exfiltration in an urban environment, what is the cyber responsibility in an urban environment hostage exfiltration? We in Air Force have been good about commanding and controlling our air assets and working with ground assets and special operation forces. But, what are some additional ideas for what you might do? For example, for the exfil team, are you hacking into the lights so that all the lights turn green magically as that exfil team goes?

Now, I'll tell you the cadets immediately think, "Well, we're going to shut down the power so that we can use our night vision goggles (NVGs) to go into the building." Then, lo and behold, there's a hospital right next to the building. You also get the opportunity to talk about ethics. You get the opportunity to talk about what would be the downside of taking out a city's power to accomplish one mission while other things are going on in that city, and what's a better tool that you might use for that. Then can you imagine it, and can you enact it?

Other things going on in here are RPA challenges—we in Air Force call drones RPAs, remotely piloted aircraft. Defense Advanced Research Projects Agency (DARPA) has sponsored several challenges both on the cyber side and on the physical side. Then at the Scowcroft Center, the Cyber 9/12. We were very proud to bring together a computer science, political science, and law team to go to that, and we were the only undergraduate institution to make it into the final round of the Cyber 9/12 Challenge. Bringing this diversity together to work through what are the challenges we're facing, what are the ways forward, is not just a technical problem, and we have to get our whole workforce moving along the lines that I've been talking about.

This is what our design thinking phases look like. You notice that there's both a month and a week. We do design sprints which are one week long. We deliver back an answer at the end of one week design sprint back to the Air Force. Then when we teach the course, we do it over the course of a semester. I can't get the cadets out of traffic or, I'm sorry, out of classes and out of their football practice and things like that, so we don't do it for one week. Instead, we spread it over the course of a semester. It ends up being about a two-and-a-half week sprint when we do it that way. These are the phases of design thinking. That last phase, that prototyping phase, is really important because that ability to try things, to try the crazy things, is what's lacking right now sometimes.

This is what our students get out of those sprints or the course. We have seen evidence along all these areas of maturity, of maturing in the officer, candidate's ability to be able to do all of the these. It is a hallmark of project-based learning that giving students a real problem and giving students a real audience to which they're going to outbrief makes those things better. It raises their game. These are not made up Jeff Collins' ideas that they're working on. These are real problems that the Pentagon has given these cadets and entrusted them to go out and help us solve.

I want to talk a little bit about cadet reactions to this. As you'd expect, the beginning of the semester, the reaction is confusion, rejection. There was no way that the cadets were going to be able to solve these problems. By the end of the semester, I came to realize I can trust the process—and we keep emphasizing trust the process. The design thinking process I will tell you is messy. None of you sitting in your chairs right now probably could draw a straight line between where you were when you started. You guys are young, so maybe you could. Those of us who are older, more experienced, we've had failures along the way. We've had benefits that we didn't anticipate happening. There's not a straight line between a starting point and an end point, a delivery point. Yet, right now at least, our acquisition systems tend to be premised upon the notion that we can get a few people in a room. We can write requirements. We can deliver those 500 pages or whatever they are of requirements to a contractor who will then deliver it possibly years later, and it will work.

The reality of the world is that the world changed the day that the requirements were written. That they were already out of date. Understanding that you're doing design and development and design and development and that it's a continuous development, operations world that we're living in is what we're pushing.

These two cadets, we use Twitter to allow our cadets to talk with people outside of the gates, and we encourage other people to help them to give them ideas about how they might solve their problems. This took place on Twitter talking about that iterative process of failure, the ability to try things to get the crazy ideas along the path to success and, as an educator, this is exactly what we want our cadets to realize, that eventually they can change the world, and, as cyber professionals, we know they have to change the world. This is their world, and they have to have the ability and the confidence and the tools to go out and change it.

Our impact in the first year, we've delivered ten projects back to the United States Air Force. We're now in two new projects for this semester. One is multi-domain command and control, which I talked to you a little bit about. The other is on smart bases. How do you enable our Air Force bases to allow our airmen to use what they're used to in their civilian life to accomplish their missions, or at least to focus more on their missions and take some of the hard parts of being an airmen out of their lives by using mobile and IoT—Internet of things technology—in a secure way.

We had both the Marines and the Navy, and we did an initial design sprint on smart bases. Now the cadets are prototyping the additional use cases to use against the architecture that the Air Force is now implementing for Air Force bases. That allows the cadets to stress test that architecture, but the use cases that they chose are also based upon cadet lives and on improving cadet lives while they're at the Air Force Academy, so they have real stakes in the outcome of their projects. That capability that we deliver back to the Air Force is focused on operational advantage, but it's also focused on building relationships with industry partners that are the driver of innovation in our society right now.

You can see at least some of the industry partners that we have worked with there—back to that cognitive diversity piece—are nonprofit, does not just recruit the big industry partners. They get small companies. They get single person limited liability companies (LLCs). They get startup companies to come in to the studio to teach us, to help us solve Air Force problems. Everybody leaves their intellectual property rights at the door. Those IP rights are protected, but the expectation in the design studio is that we're all sharing and we're not selling. What we're focused on is how do we solve this problem for our United States Air Force?

I'm really happy to see that there are so many international individuals here because this is a team sport like we have never faced before. Everybody has to be part of how we overcome the challenges that are facing us in cyber for our free world.

One of the interesting things, and this is the only success I'll go into a little detail on, is the tech transfer for the cyber risk dashboard. Three of the industry partners that came into our studio realized over the course of the design sprint that if they put their intellectual property together, they created a better solution for the Air Force. That is different. To have three defense contractors come into a studio and to realize that together they could solve this problem for the Air Force, and then to figure out with all their lawyers and all their corporate processes how to build a demonstration, to move towards a DevOps model, is, as far as I know, something we have never seen before.

Where are we going? The Air Force has committed $30 million in FY18 to build a building on the campus at the Air Force Academy. It will go right on the edge of our parade field. It will also allow us to move our fence line a little bit so that our industry partners are able to get right into our studio. Right now, they have to go through an entry control point to be able to interact with the cadets. You can see the kind of immersive labs and maker spaces that we're putting into this new studio.

Look at that list of the types of labs and maker spaces, the places for cadets to go and play, the spaces for cadets to think of an idea, to imagine an idea, to design that idea on a computer and then print it out then to stick it into an RPA or a drone to find out, did that idea work? Did that idea improve the sensor? Did that idea change the flight characteristics? That is what we need our cyber workforce to be able to do, to play with ideas, to enact ideas, to get their imagination instantiated into war fighting, if not immediately, then close to immediately.

That's what I wanted to talk to you about. That's what the Air Force is doing with CyberWorx and why we exist out there in the Air Force Academy. I don't know how many industry partners are in the room but the QR code will lead you to that nonprofit's website. That is where we announce the upcoming projects for CyberWorx. I think they wanted to see the QR code. If you could just leave that last slide up. If you're not a QR code person, then there's a URL down there at the bottom as well. With that, I think that I've got a few minutes for questions.

Q&A Segment

Audience Member 1: I have a question.

Col. Collins: Okay.

Audience Member 1: That was a great presentation. How is the nonprofit organization funded? Are there people that work there, I'm assuming?

Col. Collins: Yeah.

Audience Member 1: If so, where does the money come from?

Col. Collins: The money comes from CyberWorx. We fund the overhead of the nonprofit to do the work that we have contracted them to do for us.

Audience Member 1: The money for CyberWorx comes from the Air Force.

Col. Collins: Yes. The way that we do funding is we have different . . . They're called core function leads for different areas of Air Force operations. Cyber Superiority is one of those areas. The Cyber Superiority core function lead funds CyberWorx, and then we rely on the nonprofit, and so we pay them for their overhead services. Yes, ma'am?

Audience Member 2: Thanks. My question relates to the students in the program and how they're selected. Is this an over and above that they do while they're studying towards their degrees, and how long is the track covered? Just to put a background on that please.

Col. Collins: Like the other two service academies, we have majors at the Air Force Academy. The course, this one course, is a three-hour course that they can use as one of their academy electives. For some of the majors, it counts toward their major. They need a technology integration, either systems engineering or this management course. We're considering changing it over to a systems engineering course. We just haven't gone through the full curriculum review to decide which fits better, but it counts therefore as one of the options on their way to graduation.

Audience Member 3: You mentioned the Cyber 9/12 Challenge and that you brought an interdisciplinary team there. Can you talk about that?

Col. Collins: Talk about the challenge or the team?

Audience Member 3: And the team itself, how you brought that team together because you said they came from different disciplines.

Col. Collins: The challenge itself is . . . The Scowcroft Center and probably best if you just look for that, Cyber 9/12 Challenge; it's sort of a many graduate programs and undergraduate programs participating in that, including West Point, Annapolis—which I won't mention where they fell in the . . . Friendly fields of strife there, too, right? Not just on the gridiron.

How we bring together an interdisciplinary team, we have courses in several of our majors. For example, law has a cyber law course. Political science has courses devoted to international politics and the cyber role and that. There are professors. Then CyberWorx has . . . They're not on our books yet, but they will be coming on our books over the next year where we also have professors who are assigned to CyberWorx, but we place them into the department. The first one is a law professor, for example. We'll be hiring someone who is focused on cyber law who is able to come into our design projects, work with . . . We never know what our next projects are going to be. We're working on projects now for March and April, scheduling those. You can't say for sure that I'm going to need a cyber law expert in this sprint. We have set it up so that we've got one on staff. But, at the same time, that professor can also be teaching law courses because it's a fully-credentialed professor. It's an academic duty professional.

How we then hire the coaches. We bring a coach, a faculty member from each of those departments. The students self-select into who is able to do that. Again, the idea is we're not just looking for a computer network security major to be on that team necessarily, because we want that, but we also want a poli sci major. We also want a law . . . We don't have a law major. Forgive me. I can't quite remember what we call our law . . . It's like a minor in law. We want those different students together. I saw another hand over here I think.

Audience Member 4: Thanks. I was just curious on the student question, too. Are you working just with cadets or are you . . . I think Colorado College where I graduated, there's a partnership quad or something like that. Are you doing any work with UC Colorado Springs and so on, or is this so far–

Col. Collins: Yeah, so great question. We have a partnership with a few universities, those two that you mentioned along with UC Boulder. You saw the picture of the design studio. We just moved into that studio this summer. Part of what we're working on is the ability to bring in external teams. This idea that we could have not only a team of cadets and industry partners in studio but then we could also have a team at Carnegie Mellon or Boulder or elsewhere partnered with industry also working on the same problem, because that just expands that diversity of views.

Our student population and Carnegie Mellon student population is different in many respects. The strength of having those different ideas coming into the overall solution. That sounds easy in terms of making a student population at a different university feel like they're part of . . . ours, means that we need more technology than we have right now in terms of being able to do that, really, design integration. We have plans for doing that and we're enacting that, but we haven't pulled off that bandaid yet.

Then we also do a lot of work. We have a Department of Homeland Security (DHS). It's called the Center of Innovation which has been at the Air Force Academy for about a decade. But a lot of our focus area with them is on obviously intergovernmental solutions but then sending cadets . . . What we find a lot of value with is sending cadets to game-changing companies, especially in

Silicon Valley, where they bring back an understanding of a different culture than we have in the Air Force right now at least. But then they also bring back these ideas for technological solutions that might not occur to us if we haven't had the opportunity to send those students out.

The reason that I bring this up during an answer to your question is we are working on . . . DHS increased our funding for that which I appreciate. That enables us to also work with ROTC programs to send some of those cadets out as well, and so we're working through how . . . We do those agreements with those industry partners on . . . They're called cooperative research and develop agreements, CRADAs. We're working through how do we do that to ROTC, for ROTC, but then also sister service academies. We would love to have a West Point and an Air Force Academy cadet, for example, PERG.

We had a cadet that we sent to Facebook last summer who came back. The project that he's now working on after being at Facebook is an AI bot that is a Facebook cutting edge technology, but he's using it for his . . . He's the captain of the golf team. At the end of practice, he wants to know do I need to hurry back to eat at Mitchell Hall, which is the big cafeteria, or should I just take my time so that I can eat off base? That is a cadet problem. But, the benefit for the Air Force is that I now have a cadet using AI technology and, oh, by the way, he's got 400 other cadets now using this AI bot, understanding the cyber security implications of an AI bot. Then I don't know what problem he's going to face as a second lieutenant that this AI bot might be a solution for. But what I like is that I've now got a future airman who is willing to think and apply and figure out how to make something work.

That is cyber workforce. We need that talent to be able to apply imagination no matter what the situation is. Other questions? Yes, sir.

Audience Member 5: I'll do sound check.

Col. Collins: Okay. Command voice.

Audience Member 5: What did you learn about acquisitions and authorities that might apply as lessons to the broader problems of acquisitions in the government?

Col. Collins: That's a great question. We have a sprint coming up. To answer that, when I was the chief staff up on working with the CIO staff, the biggest thing is colors of money. In the United States, we have a . . . We call them colors of money but they're appropriations by Congress that determine how you can deliver capabilities, and how you do that depends on the color of money.

The question that keeps coming up for is, well, look, are you doing design, or are you doing development? Because that determines which color of money under the old system you should be using. That's a bit of a nonsensical question in a DevOps environment because you are always doing design. You are always doing development, and so barriers like that . . . I talk about it in terms of teaching new muscle movements. We've been teaching a lot of new muscle movements and getting so the purpose of the design sprint is that we now have evidence-based complaints. We thought it was going to be a problem. It turns out it is a problem. We have evidence that it's a problem. Now, let's go talk to the policymakers in the Pentagon and, if necessary, the LL, legislative liaison, to talk about how we overcome those barriers.

A lot of times we find that it's a policy that is based on a reading of a law that might not have been the intent. We use hashtags for all of our design sprint. The hashtag for that is Gumby stacks, that idea of more flexibility in a way that we use money. In CyberWorx fashion, we'll do a one week sprint, and we'll come up with twenty . . . Usually when it necks down, you've got five or six areas that we can see forward movement. Of course, our acquisition professionals in the Pentagon are very interested in hearing that and seeing what we can do.

Presentation Slides

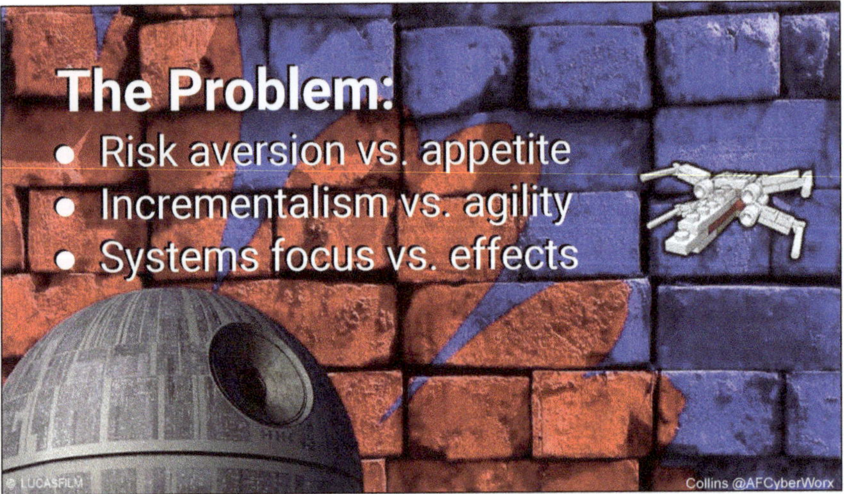

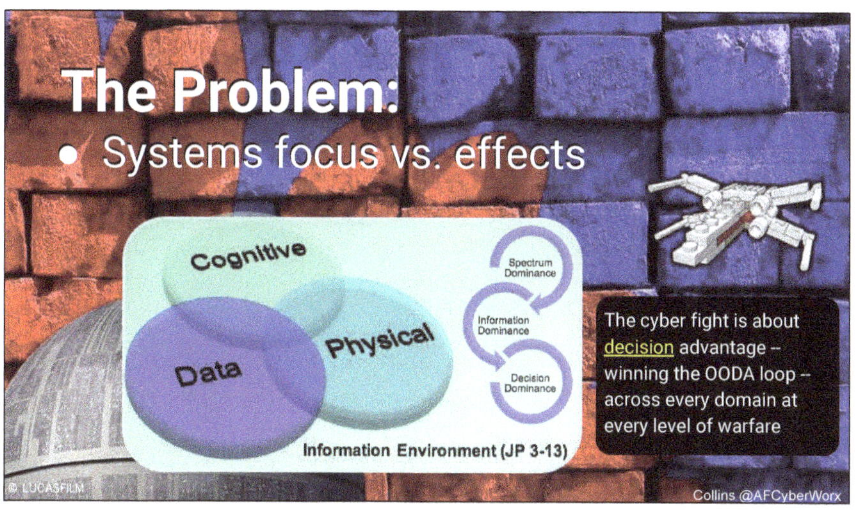
The Problem:
• Systems focus vs. effects
Cognitive
Data
Physical
Spectrum Dominance
Information Dominance
Decision Dominance
Information Environment (JP 3-13)
The cyber fight is about decision advantage -- winning the OODA loop -- across every domain at every level of warfare
© LUCASFILM
Collins @AFCyberWorx

AF CyberWorx Vision:
Unlocking the power of people to unleash the power of cyberspace for America's Air Force
Collins @AFCyberWorx

AF CyberWorx Mission:
Accelerate operational advantages by moving toward a simpler, more intuitive and agile Air Force
Partner
Innovate
Educate
Build
Collins @AFCyberWorx

CYBERWORX
Collins @AFCyberWorx

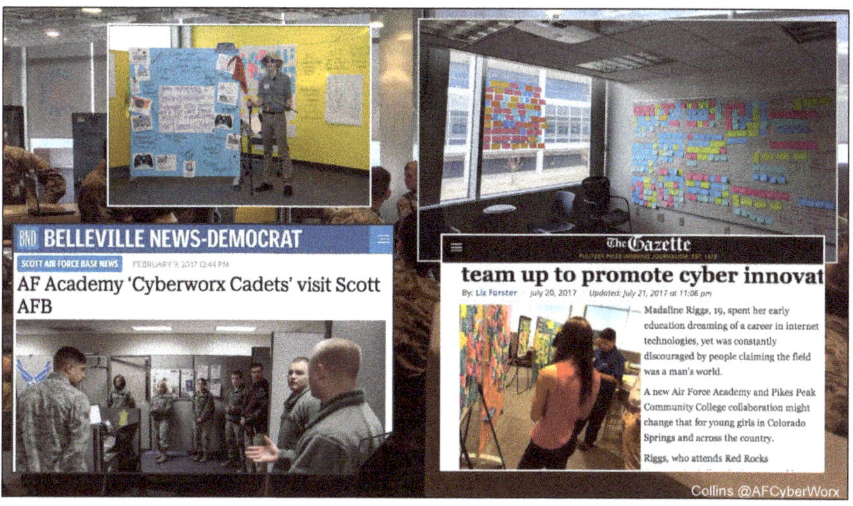
BND BELLEVILLE NEWS-DEMOCRAT
SCOTT AIR FORCE BASE NEWS
AF Academy 'Cyberworx Cadets' visit Scott AFB
The Gazette
team up to promote cyber innovat
Madaline Riggs, 19, spent her early education dreaming of a career in internet technologies, yet was constantly discouraged by people claiming the field was a man's world.
A new Air Force Academy and Pikes Peak Community College collaboration might change that for young girls in Colorado Springs and across the country.
Riggs, who attends Red Rocks
Collins @AFCyberWorx

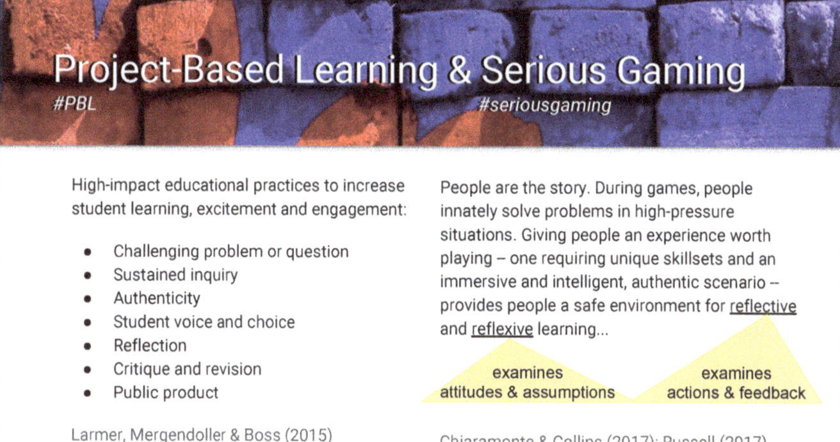
Project-Based Learning & Serious Gaming
#PBL
#seriousgaming
High-impact educational practices to increase student learning, excitement and engagement:
Challenging problem or question
Sustained inquiry
Authenticity
Student voice and choice
Reflection
Critique and revision
Public product
Larmer, Mergendoller & Boss (2015)
People are the story. During games, people innately solve problems in high-pressure situations. Giving people an experience worth playing – one requiring unique skillsets and an immersive and intelligent, authentic scenario -- provides people a safe environment for reflective and reflexive learning...
examines attitudes & assumptions
examines actions & feedback
Chiaramonte & Collins (2017); Russell (2017)
Collins @AFCyberWorx

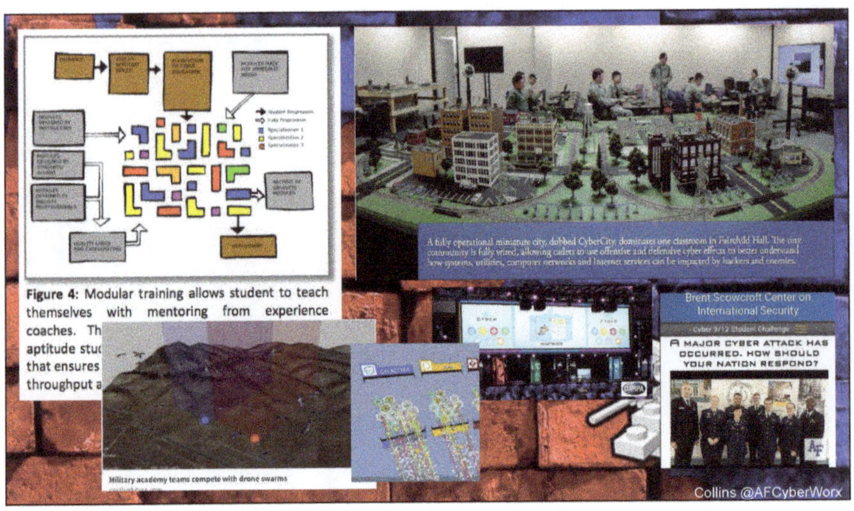
Figure 4: Modular training allows student to teach themselves with mentoring from experience coaches. Th
aptitude stu
that ensures
throughput a
A fully operational miniature city, dubbed CyberCity, dominates one classroom in Fairchild Hall. The city community is fully wired, allowing cadets to use offensive and defensive cyber effects to better understand how systems, utilities, computer networks and Internet services can be impacted by hackers and enemies.
Brent Scowcroft Center on International Security
Cyber 9/12 Student Challenge
A MAJOR CYBER ATTACK HAS OCCURRED. HOW SHOULD YOUR NATION RESPOND?
Military academy teams compete with drone swarms
Collins @AFCyberWorx

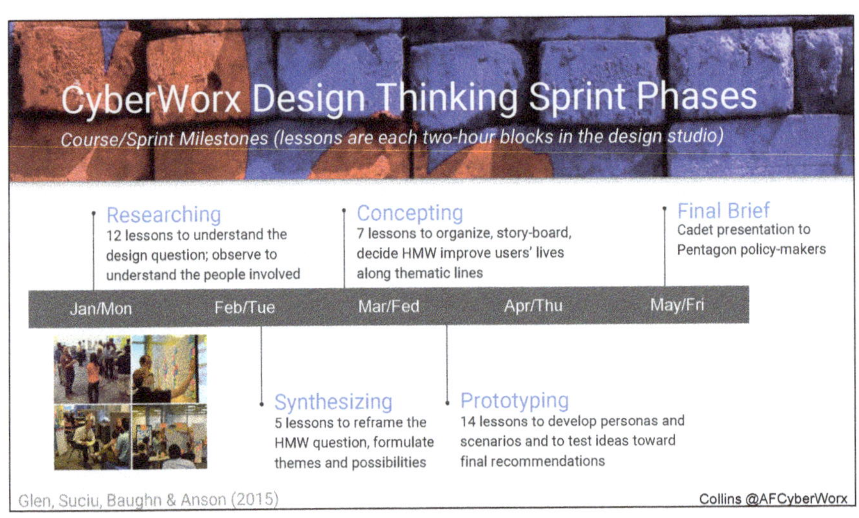
CyberWorx Design Thinking Sprint Phases
Course/Sprint Milestones (lessons are each two-hour blocks in the design studio)
Researching
12 lessons to understand the design question; observe to understand the people involved
Concepting
7 lessons to organize, story-board, decide HMW improve users' lives along thematic lines
Final Brief
Cadet presentation to Pentagon policy-makers
Jan/Mon
Feb/Tue
Mar/Fed
Apr/Thu
May/Fri
Synthesizing
5 lessons to reframe the HMW question, formulate themes and possibilities
Prototyping
14 lessons to develop personas and scenarios and to test ideas toward final recommendations
Glen, Suciu, Baughn & Anson (2015)
Collins @AFCyberWorx

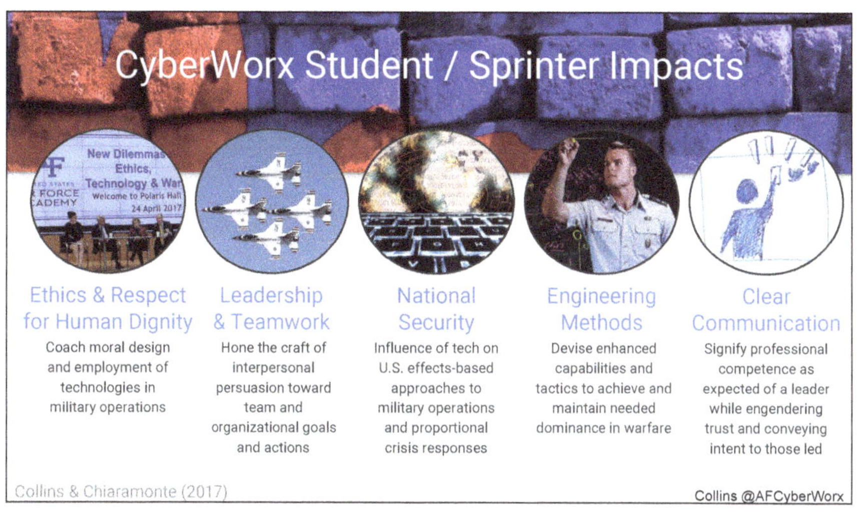
CyberWorx Student / Sprinter Impacts
New Dilemmas
Ethics,
Technology & War
Welcome to Polaris Hall
24 April 2017
Ethics & Respect for Human Dignity
Coach moral design and employment of technologies in military operations
Leadership & Teamwork
Hone the craft of interpersonal persuasion toward team and organizational goals and actions
National Security
Influence of tech on U.S. effects-based approaches to military operations and proportional crisis responses
Engineering Methods
Devise enhanced capabilities and tactics to achieve and maintain needed dominance in warfare
Clear Communication
Signify professional competence as expected of a leader while engendering trust and conveying intent to those led
Collins & Chiaramonte (2017)
Collins @AFCyberWorx

At the outset my reaction was confusion and rejection. I believed there was no possible way a group of cadets could analyze the problem. I questioned the class's probability of success...
Quotation is NOT attributed to the cadet pictured
Collins @AFCyberWorx

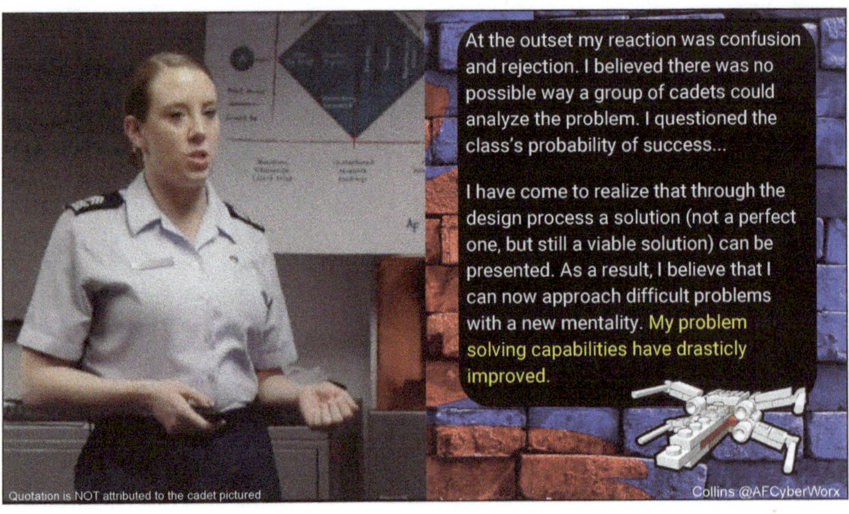
At the outset my reaction was confusion and rejection. I believed there was no possible way a group of cadets could analyze the problem. I questioned the class's probability of success...
I have come to realize that through the design process a solution (not a perfect one, but still a viable solution) can be presented. As a result, I believe that I can now approach difficult problems with a new mentality. My problem solving capabilities have drasticly improved.
Quotation is NOT attributed to the cadet pictured
Collins @AFCyberWorx

@[student] hit this nail on the head - an iterative process of failure is part of the path to success...
Quotation is NOT attributed to the cadets pictured
Collins @AFCyberWorx

@[student] hit this nail on the head - an iterative process of failure is part of the path to success...
As long as you keep designing and innovating, failure and setbacks are only temporary. Eventually, we can change the world.
Quotation is NOT attributed to the cadets pictured
Collins @AFCyberWorx

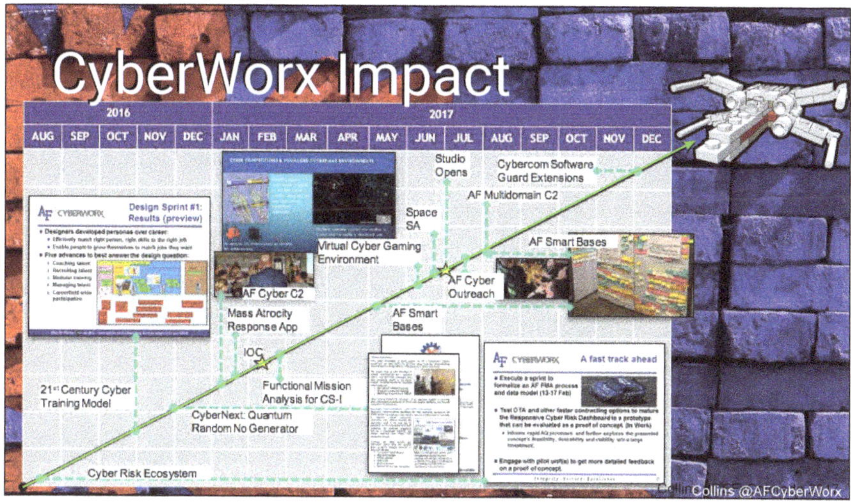
CyberWorx Impact
2016
2017
AUG
SEP
OCT
NOV
DEC
JAN
FEB
MAR
APR
MAY
JUN
JUL
AUG
SEP
OCT
NOV
DEC
Studio Opens
Cybercom Software Guard Extensions
AF Multidomain C2
Space SA
Virtual Cyber Gaming Environment
AF Smart Bases
AF Cyber C2
AF Cyber Outreach
Mass Atrocity Response App
AF Smart Bases
IOC
21st Century Cyber Training Model
Functional Mission Analysis for CS-I
CyberNext: Quantum Random No Generator
Cyber Risk Ecosystem
Design Sprint #1: Results (preview)
A fast track ahead
Collins @AFCyberWorx

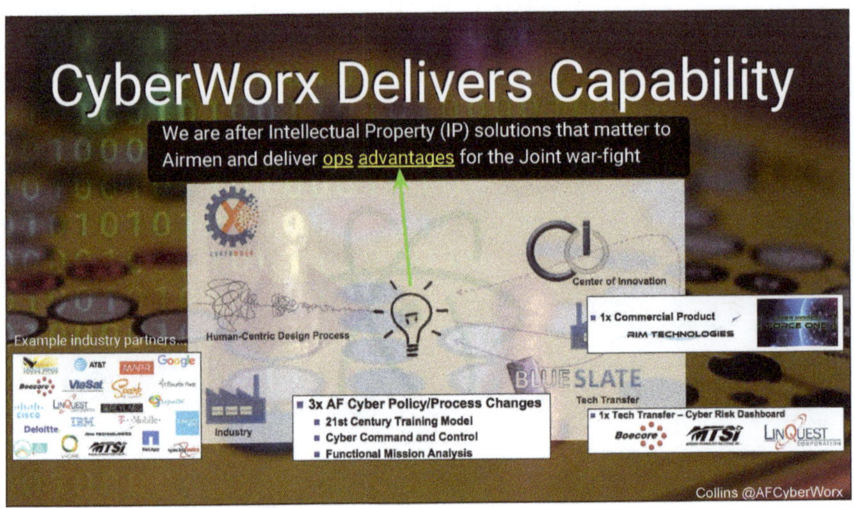
CyberWorx Delivers Capability
We are after Intellectual Property (IP) solutions that matter to Airmen and deliver ops advantages for the Joint war-fight
Center of Innovation
Human-Centric Design Process
Industry
Example industry partners...
1x Commercial Product
RIM TECHNOLOGIES
BLUE SLATE
Tech Transfer
3x AF Cyber Policy/Process Changes
21st Century Training Model
Cyber Command and Control
Functional Mission Analysis
1x Tech Transfer – Cyber Risk Dashboard
Collins @AFCyberWorx

CyberWorx Future
Military Construction (MILCON)
FY18 Air Force CyberWorx (33K SF) $30.0M
Donor/Gift Funds Construction & Equipment
FY18 Air Force CyberWorx Center ~$25.0M
- Public-Private ~10K capacity boost (~$17.5M)
- Building finish/tech enhancements (~$5.0M)
- Immersive Labs & Maker Spaces (~$2.5M)
Immersive Labs/Maker Spaces:
Cybersecurity
Forensics & Reverse Engr
Cyber Training x2
Cyber Law, Ethics, Policy & Digital Humanities
Telecomms & RF
Industrial Controls
Next-Gen C2
Robotics & Autonomous Systems
Stillman Parade Field
Design Concepts (10 Project Capacity)
Collins @AFCyberWorx

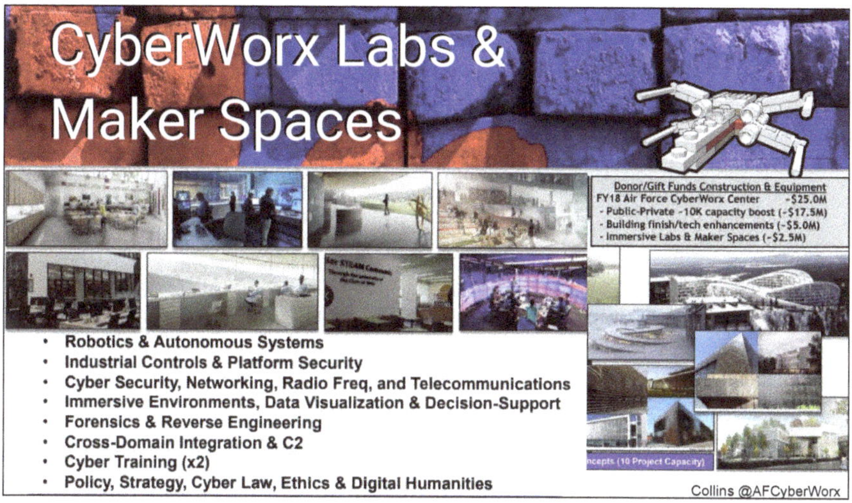
CyberWorx Labs & Maker Spaces
Donor/Gift Funds Construction & Equipment
FY18 Air Force CyberWorx Center ~$25.0M
- Public-Private ~10K capacity boost (~$17.5M)
- Building finish/tech enhancements (~$5.0M)
- Immersive Labs & Maker Spaces (~$2.5M)
• Robotics & Autonomous Systems
• Industrial Controls & Platform Security
• Cyber Security, Networking, Radio Freq, and Telecommunications
• Immersive Environments, Data Visualization & Decision-Support
• Forensics & Reverse Engineering
• Cross-Domain Integration & C2
• Cyber Training (x2)
• Policy, Strategy, Cyber Law, Ethics & Digital Humanities
ncepts (10 Project Capacity)
Collins @AFCyberWorx

Air Force CyberWorx: Projects & Game-Based Approaches to Cyber
GO AIR FORCE!
GO AIR FORCE!
Jeff Collins @JeffCollinZ
Director @AFCyberWorx
UNC ILSS Symposium, 9 November 2017
Upcoming projects: www.c-trac.org/cyberworx
CYBERWORX

9

Colonel Laszlo Kovacs

As presented at the 2017 Civil-Military Symposium
Hosted by the Institute for Leadership and Strategic Studies
University of North Georgia

Thank you very much, Dr. Wells, for the kind introduction. I'm very glad to be here. It's a good opportunity, it's a great opportunity, to share our activities mainly on the cyber fields. But I really know that I'm between you and the end of the symposium. That's why I'll try to keep to the time.

So, yesterday we started the panel meeting to share our activities, to share the information of our activities in the cyber field. Dr. Krasznay introduced our flagship on the cyber security. This is the cyber security academy in the National University of Public Service (NUPS), which is not a brand-new university in Hungary; we have a five- or five-and-a-half years' old history. It goes back into the past. We have a 200-year-old history because the military higher education started 200 years ago in Hungary within the Austrian-Hungary empire, of course.

So, we are from Hungary. Just some basic information about Hungary. As you know the country is an Eastern European country. It's a relatively small country with only 10 million peoples. The gross national domestic product (GDP) is not so high. It's takes the sixty-first place among the countries. Our neighboring countries are mostly part of the European Union and the North Atlantic Treaty Organization (NATO). There are two exceptions: Serbia and Ukraine. That's why we have to take into consideration a lot of things in the cyber sphere, cyber space, because Serbia is very, very close to the European Union. They have cooperation with the European Union, but in Ukraine there is war.

We can characterize it as a hybrid war, but, to be honest, it's a war. It's just 500 kilometers from our neighboring borders. That's why we have to look into the rations and the ration influence on this region, but we have to be careful because Hungary mostly depends on the Russian nature of gas and other economic relations that we have. That's why I will refer later a little bit more on how can we see in the cyber sphere and which kinds of things we have to take into consideration during our research.

As concerns the military, the military is only 30,000 people, so it's a little bit not too big, let's say. It's a little bit less than the US air force or the US army. We have only two services or two branches: the ground forces and the armed forces. We have professional armed forces, with no conscription since 2004. The expenditure is not so high. It's only 1% of the GDP in the last year, but there is a development plan for the next ten years to make it twice more in size until 2026. Last Friday, the minister of defense was at our university and made a brilliant speech about the new development plan, and he mentioned a lot of new things that came up now, which kind of things that we have to develop and will develop. He emphasized and underlined the first line, that the cyber capabilities of Hungary or the Hungarian armed forces would be the first that we have to develop in the future.

We have our new university: the NUPS. We love the acronyms. This the National University of Public Service. This is the main building. To be honest, it's the oldest military academics institute among our countries. As I mentioned, we were part of the Austrian-Hungarian empire, where there were only five military academies at that time, and we were very proud—and we have to be very proud—because the Hungary military academy was the only academy that was allowed to teach not in German but in

Hungarian. It was unique. Today we have a very unique university since the beginning of 2012, because the Hungarian parliament of the Hungarian government made a decision to merge the National Defense University, the Police Academy, and the faculty of the most famous Hungarian university, the Public Administration faculty of the Corvinus University.

We have a new university from that time onward. We started with three main faculty at the beginning of 2012, the faculty of military as we call official military sciences officer training; faculty of law enforcement; and faculty of public administration. In the beginning of 2015, a new faculty emerged. This is the faculty of international and European studies, and in the beginning of this year, we got a new faculty that was mentioned yesterday, the faculty of water sciences.

It could be very interesting because we have a new university, and the philosophy behind the new university is to train together the full range of public service workers for the future. It doesn't matter that they are military or police or public servants for the future, and in that regard it would be very strange that we are also charged with the water sciences. We have two main rivers and very often very, very high floods in Hungary. That's why we need to deal with the water and the water sciences.

However, the water or handling the water could be a critical infrastructure. That's why we have to put into our research focus on the water sciences, or on the water infrastructure as critical infrastructure.

Just some basic facts about our university. As I mentioned, it's a state university. The main location, the HQ, is in Budapest in the center of the city. It's a very beautiful city. To be honest, it's a medium-sized university in Hungary because we have 4,000 regular students. I know that it could be a small university in the meaning of the US or the other bigger countries. We have more than 70,000 students every year who are participating in further trainings. We have five faculties as I mentioned. We have more than 1,000 professor instructors and admin workers. We have a very centralized management. It means that we have, let's say, a supervising committee, because we are operating not under the ministry of education, higher education, but we are operating under the minister of defence, minister of interior, minister of justice, and the ministy of the prime minister's office. They are four ministers. It's half of the Hungarian government. Could you imagine our situation every day that we have to solve and we have to fulfill?

We have a new campus, because two years ago the government decided to rebuild the old military academic building that I showed you. Here is the new campus surrounding a beautiful park; a lot of new buildings emerged here for the classrooms. We have new accommodation facilities. We have a new sports field over there, and we have new facilities for the police forces training. Just a picture about the beautiful new campus. We are very proud of it, let's say, because during the last thirty years, there was no such kind of development for this kind of university or college. That's why it's a huge step forward for us towards serving the country, or serving the nation as our motto says.

This is the main building entrance. We have, as I mentioned, new accommodations facility. This is the dormitory for the civilians. We have new lecture rooms, board rooms or lecture hall, and smaller lecture rooms, but the development will be continued because there is a huge park, as I mentioned, over there, and that will be our new sport facilities. This is my favorite; there will be a horse riding area in the center of the city, because it's compulsory for the police officers.

The military faculty has bachelor's, master's and PhD degrees levels of education. In the bachelor's degree program, we have very career training oriented programs. We have a leader and commander training program. We have a military logistics. We have maintenance. It may sound very strange, but this is its official name.

It's integrated; the signals, the IT, the electronic warfare, and the aviation engineering experts are the future experts. We have one new program; this is the aviation program, where the new pilots we trained and the new ATC, air traffic controllers, will train from the next year on.

We have master's degree programs. They are similar, but we have a defense IT program for the civilians—who are the civilian guys who are involved in the defence sector. Besides the university programs, we have a lot of courses, because we are the only higher education institute in Hungary who can, and who are allowed to, train military and police. There is no other police; there is no other military university or colleges. We are the only one. That's why the Ministry of Defense (MOD) focuses on the other courses, the other further trainings from us, as written in the slides here.

In the next slides, there is the main philosophy behind the university. We have fourteen main modules which are compulsory for everybody. It doesn't matter that it's police cadets or military cadets or civilian students. They know every kind of information—I mean basic information about the police, the military, the public administration, the constitutional law, the processes of the state, and every kind of thing that they need. They do it together. As I mentioned, we have a lot of disasters, and this is the most beautiful example of our philosophy. How can they do it together in the future? Because they learn together, these subjects, they will know each other, and they will handle the future problems together.

In addition, we have a joint exercise every year when they have to prove their knowledge in this field, in these areas. As I mentioned, we have bachelor's, master's, and PhD programs. We have four doctoral schools, PhD schools; two of them belong to the military faculty. These are the doctoral school of military sciences and the doctoral school of military engineering. The two PhD schools offer a lot of research topics for cyber defense and cyber security. We have to divide it into two parts, cyber defense and cyber security, because the EU, the European Union, wants to use cyber security and not cyber defense—instead of cyber defense—that's why we use cyber security. In NATO we use cyber defense as a common terminology. That's why we have both of them in this course.

We have a generalist program, of course. It's almost twenty years old. It's an almost one-year long generalist program of courses taught in the Hungarian language, but it's an international course because we have Chinese, sometimes we have Russians, Germans, Italians—lieutenant colonel and colonels—in the program. Maybe they will be the new military attaches in Hungary. That's why it's a good opportunity to see how the Hungarian defense sector works, and so on.

We have a lot of research and development works. Just a few examples: in the beginning of 2000, we started to research Unmanned Air Vehicles (UAVs). We had a proposal for the European Union within a framework of the (Fifth Framework Programme) FP5. This is an initiative of the European Union; it was founded by the European Union. For our proposal, we wanted to try to figure out for which kinds of civilian tasks should we use the UAVs. It was very unique at that time because the armed forces, maybe the armed forces or the land forces, used the UAVs at the time, but there was no civilian examples for that. We have under the research activities that it's a biological container. It's brand new that we made that. We have a lot of electronic warfare assets that we made for the armed forces or that we can offer the armed forces. Without a library, we cannot do any research. That's why we are very proud of our library. It includes more than 600,000 books, and 1,000 or more in the international database, the online database—and students and researchers can use it.

We have a lot of international cooperation that I mentioned yesterday because we realized that alone we cannot be successful in the future or in the present. We participate in the central European forum of military education and, as Dr. Vaus mentioned, we made a consortium just a couple of years ago. This is the International Military Academic (IMF) forum. The IMF's main aim is to establish or set up at minimum one semester, one international semester, for the bachelor's degree training all around Europe. In the EU, in the twenty-eight countries in the EU, we have fifty-eight different military academies and military universities with fifty-eight different curriculum and different training systems. That's why we need to harmonize at minimum in one semester. It's done. We will finish this at the end of this November, and we will see for the future. It should be published on the European Union or the European Commission web pages.

We concentrate on a lot of things, and we try to include in this semester the electronic warfare and the cyber security. I will refer a little bit later to how can we do it, because we tested it, and we've got a lot of very interesting feedback from the students from the international environment on how they look into electronic warfare today as well as cyber warfare.

Just some picture about these activities. We have multinational exercises. These are part of the education. This is a logistic exercise with the British logistics school. We have one of their exercises. It's also our logistic exercise. This is another. It's Austrians, Czechs, and Hungarians participating. We have communication exercises with the French communication school, and we focus on cyber security and cyber defense.

I have to emphasize that one of the key issues in the NUPS is to strengthen cyber security all around the country, mainly in the public administration. The public administration means the police, the military, and the public service, from the higher level of political decision makers to the bottom line, to the municipalities' workers, let's say.

This is the second thing that we have to mention here. We started in the beginning of the '90s or a little bit earlier because we started in the '70s, or the predecessors started with electronic warfare. I'm sure that you know at the time Hungary was a part of the Warsaw Pact. Within the Warsaw Pact, electronic warfare was the key issue among the alliance. Every country, every member country has to build up or had to build up very, very massive electronic warfare components and very, very sophisticated, let's say, sophisticated electronic warfare capabilities. Of course, based on Russian technology. That's a very, very huge heritage and a very, very hard heritage for the '90s because in 1990 when the political regime changed, we got the old Russian technology. It was usable, of course, but it was very analog. It's a little bit obsolete, and there was no money to change into the new environment. That's why it was very, very hard.

After that, in the middle of the '90s, we started to research and started, let's say, to do it, to manage the information warfare and information operations. When the NUPS came up, we tried to establish a cyber defense think tank based on the experiences of the electronic warfare, the information warfare, information operations. After that, we planned a cyber research center, and we did a lot of research work within the scope of cyber security and critical infrastructure protection, mainly the critical information infrastructure protection. Meanwhile, we tried to integrate the cyber security into the teaching program or the training programs because every bachelor's degree student or cadet has to know the basic things about cyber security. It doesn't matter that the student may in the future become a police officer or a member of the military or a public servant.

As I mentioned, in the master's degree level, we have a different and separate defense IT graduate program, but the problem with this graduate program, this defense IT graduate program, is we have subjects in the other master's programs. I know that our system is a little bit different than the US system. We have bachelor's, master's, and PhDs, but they are definite training programs; there is no majority, there is no minority, just the definite subjects within the training programs.

As the cyber security related field is constant, we focus on cyber strategies because just a couple of years ago there was no cyber strategy for Hungary. There was no cyber security strategy even in the EU, in the European Union. That's why we had to figure out how could we establish or how could we set up within a very rapidly-changing international environment a new strategy, or how could we fund the basic elements for our cyber strategy?

Problem with this research is we had some cyber warfare researches. It's a huge challenge, to be honest. As I mentioned, we have a very huge dependency on the Russians, in our economy and in the nature of gas. The 99% of the nature of gas, let's say it, 99% came from the Russia. That's why we have to be very, very careful with these kinds of things. We deal with cyber terrorism because it's not a thing of the future, it's a reality now. Unfortunately, the EU cannot do anything against cyber terrorism. It started much earlier, before the IS. Much earlier than the Al Qaeda. There were a lot of European terrorism organizations. They used the internet. They used social media. Well before IS and before Al Qaeda.

As I mentioned, we deal with the methods and weapons of cyber warfare, and we did a lot of research on critical information infrastructure protection. We focused on the warnings and the situation awareness. We made a scenario like the digital Pearl Harbor with Dr. Kraznay. It was a digital version of the Battle of Mohacs because Mohacs was the same situation, the same tragic point in Hungarian history as Pearl Harbor was in American history. In the beginning of the sixteenth century, the Turkish Empire defeated the Hungarian kingdom's armed forces, and it was very suddenly and very sad. Temporarily, they occupied the country. It meant that for 150 years they were located in Hungary. That's why we called it the digital Mohacs.

We tried to figure out the main points of the infrastructure of Hungary that we can destroy and all kinds of things, from the power grid to the public administration, that we can cut off. We published this scenario at a hacker conference. There were 600 guys, very bright guys, in the audience. After the presentation, they came to us and said, "Sir, we have other ideas. Please do it in this scenario." We mentioned that, thank you very much. It wasn't my goal or our goal. Our goal was very simple: to focus the political decision makers' attention on the problem that we have no laws on cyber security. There was no law at that time on the critical infrastructure. It was eight years ago. After that, in just three years, we've got a cyber security strategy; we have a particular law on critical infrastructure, mainly on the critical information infrastructures.

We have these activities; within these activities are a lot of connections and cooperations with the MOD, the ministry of interior, and within the scope of cyber security, we have a very good connection with the National Cyber Defense Institutes. This is the host of the government authority in Hungary based on one of the national security agencies. We have a lot of connections with the other universities and think tanks, not only in Hungary but all around Europe.

We have a very good connection with NATO cyber security or the Cyber Defense Center of Excellence. We participated in, and we've got the chance to participate in the lock cyber exercise. That's why we propose to UNG that in the next spring, we would like to set up a common theme to do it, and for the CC, we made a survey on the Hungarian defense organization just two years ago.

We have a lot of European Security and Defense College (ESDC) workshops in Hungary because the ESDC is the virtual college of the European Union in the defense sector. They execute a lot of exercise and a lot of workshops. That's why we involve them, and we made common things to do during the last years. Within the IMF, as I mentioned, we tried to focus on cyber security, and we try to involve cyber security in the common international semester for higher military education.

It was amazing to see how the cadets get the knowledge if it compared with electronic warfare, because they didn't understand electronic warfare. It was amazing, the reason why we need electronic warfare, because we have GPS, we have computers, we have networks, we have radios, and we have everything. It's sure that we have to protect it, but what about electronic warfare? That's the main thing that we have to explain over and over, that we need, I mean in the armed forces, we need electronic warfare. But we have to be careful, because we have to learn the lessons from the Ukrainian war because the Russians or a Russian-supported somebody used electronic warfare against civilians. That's why we have to use electronic warfare for the knowledge of the law enforcement units as well. That's the huge challenge that we are facing now.

So, we try to figure out some reference curriculum for cyber security training. It's on the table of the European Union. We will see the future, but we hope that it could be a reference model or we change a little bit, but it could be a reference model for cyber security.

This is the other flagship that I mentioned. This is the cyber security resource center that I'm the head of. It's a two-year research program, so it terminates. It's a very small team, just a couple of guys in this research center. We have fifteen researchers. I have seven professors and eight PhD candidates. The budget is not so much, not so high; it's $300,000 for the next two years, and we have to figure out six research pillars. The main aims are to coordinate the cyber-related research activities within the NUPS, and we have to double up a lot of materials, teaching materials, for the new cyber security academy that was mentioned yesterday.

The other main aims are, bottom line, that we have to double up the cyber security and the cyber awareness within the public administration in Hungary because the money came from the

European Union, and the main aim of the European Union is to change or develop the whole public service in Hungary, because we need it. I'm sure that you know that the European internet penetration rate is about 80%, so in Hungary it's much higher, but we have a lot of things that we have to develop, for example, the e-government solutions, because the people don't want to use it somehow. We have to force them to use it because it's much easier, but we have to make the security e-government solution for them. That's what we have to focus on as well.

This is the research model for our processes. We try to, at the bottom line, we try to establish some smart environment, and, based on the smart environment, we try to get the awareness, or to develop the awareness, the information security or, let's say, cyber security. However, we have to convince the military and the civilian decision makers that we need it. That's why we need some leadership, and at the highest level, we need new strategy in this field.

As I mentioned, we have six research pillars within this research center, and we started some work here with cyber security awareness. It means that we define different groups. It puts them into the labs, and we try to measure the security, the cyber security awareness. We've got some knowledge, but what they are doing, what they are thinking about, is the cyber security, how they use the computers, the smart phones, and on other solutions. After that they trained them an indeterminate time; we've got some training courses for them, and after that, we measure them again, and we compare the inputs and the outputs. We can change the two things, if it's needed. Is it usable, this kind of training? Because most of the training, how can I say it in a politically correct way, doesn't work properly, let's say it. It is very, very expensive, but the training is not efficient, let's say it. That's why we need to improve the training, and via the training or through the training, we have to improve cyber security awareness within the public sphere or the public sector.

In the second pillar are the smart cities. It's a huge problem because 75% of the population live in an urban area, in cities. We have to make an efficient or effective services of them. That's why we need to avoid the security issues of smart cities. In addition, the Hungarian government has a plan to focus on their development without addressing any security issues. That's why we have to provide the security issues for that.

The fourth pillar of the research model are the challenges, the cyber security challenges for organizations. It's not at the tactical level. At the strategic level, we tried to influence the decision makers, not only for the budget that we need more money, but for how they're thinking about it in the future, how can they make a decision about further strategies?

After that, the fifth pillar are the real cyber strategist researches, because we need new strategies for the armed forces, and we need new strategies for the civilian sector, cyber security strategies, but we have to avoid the cyber warfare strategy a bit because it's missing in Hungary. It came from the European Union, I mean the money, so we cannot use it directly for the armed forces or for military purposes. That's why we have to shift a little bit in the future.

The sixth pillar are cyber crime and IT forensics. This is the weakest point, to be honest, frankly speaking, because cyber crime is emerging during the recent years. Unfortunately, we can see the eastern countries' influence in Hungary, and we can see that nobody can handle it within this region. In addition to the IT forensics, it's very, very expensive. If we try to do something in the labs or in the university, it's amazing how expensive these things are.

The results will be the new training programs, as I mentioned, and the new material for the training, not for the country but in the international society. That's about the future, and I think the slides can be summarized on the two amazing days that we had here because we have a lot of open questions from the technical and human resources sides, because in the public service we cannot guarantee such a high amount of budget or salary for these guys. It doesn't mean just the technical guys, but the IT experts or the security guys,as well, because the private companies or the private sector can give them twice as much or much higher salaries. I think it's the same all around the world. We have to handle it.

The cyber crime I mentioned. The cyber terrorism, unfortunately, is in Europe. We can see a lot of radical web pages, a lot of radical newspapers on the internet. They are very simple, but they can influence the guys who mainly can be influenced, unfortunately. We can see not every week, fortunately, but every month or every other month serious terrorist attacks all around Europe. That's why we have to handle the cyber terrorism question.

The cyber influence and the cyber deterrence, these are the two questions that we have to handle in a very, very politically correct way, let's say it, because of the eastern countries that I mentioned. There are a lot of technical questions. The cloud computing or the quantum computing, especially the quantum computing, if it would have a breakthrough within the next years, we can throw away the cryptography that we use, for example, for our banking methods or bank transferring. There are a lot of signs in the quantum computing, which is much faster. I'm sure that you know a lot of things about quantum computing. It's amazing, how the new era of the IT will look in the near future. If as I mentioned there will be a breakthrough that comes into effect within the next five, ten, or fifty years—we cannot predict the turning or the break point. We can throw away the cryptography that we use today. And we have a lot of other questions that we mentioned here during the last two days.

Thank you very much for your attention. These are our activities on cyber security and cyber defense at NUPS. Thank you very much once again for your time.

I'm ready for the questions, but be very polite to me because of my English that I mentioned yesterday. Yes sir.

Q&A Segment

Stollard: Thank you very much for the talk. I'm Christian Stollard. I've done some studies of various European countries. They're all facing the same series of problems. Some of which don't operate 24/7; they've all got different doctrines and strategies and so on. One thing that strikes me is this could be discussed more at the European level, at the EU level. Organizations like Africa, Near East & South Asia (ANESA) could be bolstered and strengthened; Europol, Interpol, these kinds of organizations could be strengthened. NATO itself at a military level could do more, and through that you could leverage some intelligence agency support as well, or increased information and intelligence sharing, which would help you and help many other nations across your Atlantic area, especially the smaller nations with less resources to plow in.

Col. Kovacs: Thank you very much for the question. I think it's very complex, I mean the answer, because during the last years, mainly during the last five years, the European Union took a lot of steps toward reaching a higher level of cyber security. For example, the Europol—this is the common police for the European Union—made a European cyber-crime center, and they have gotten a lot of success during the last years. They covered a lot of cyber crime. It means that more than 400 billion euros was recovered that was stolen by hackers, or let's say by the criminals. This is the most valuable example, I think.

The other example for which the European Union has initiative research and development project or framework is the Horizon 2020. It's got a huge amount of money. It's got more than 80 billion euros for the six years, which will use or should be used for the kind of development that I mentioned here. As an expert, let's say an expert, a valuation expert, I saw many, many very, very valuable projects within the Horizon 2020 with the main aim to develop cyber security, not only in the public sphere or the public sector, but in the private sector as well.

The other question is the military. How does NATO handle the kind of threats that we are facing now? Unfortunately, I couldn't give you an exact answer yet or now, but NATO was the first international organization which could guarantee 24/7 capabilities at the end of 2012. The European Union in this year established that for the EU only. That's why we can state that the NATO countries, not only in Europe but here as well, can guarantee the military systems, I mean the military networks security in terms of incidents response or incident handling. NATO has a lot of research work.

As I mentioned, the NATO Cyber Defense Center of Excellence is developing in an emerging way, I think. The CyCon conference in this year, at the beginning of June, was the best example. How can they collect the common knowledge to develop within not only NATO but in the civilian sphere as well, collect the knowledge for future cyber security? During this week, there was the CyCon in Washington, the CyCon in the US, because there is a huge interest within NATO or in other countries in how can we do it together? Because, for example, the cyber terrorism or the cyber warfare cannot be done alone.

On the other hand, information sharing I think is the backbone of these things. It's very, very sensitive. We saw within the terrorism issue that it sometimes, somehow it doesn't work properly because the big country, or the big countries, let's say it, here in the US, the big countries have the capabilities and the technology background to it, but what about the small countries? How can they collect the information? How can they make proper reconnaissance and other things like that? Unfortunately, the smaller countries have no money to develop ISR capabilities. In a limited way, yes, we have, but we have to use it, or we can use it, only in the military and not in the national security sphere. That's why I mentioned that it's very, very complex, and I cannot give you a certain answer from here. Thank you very much.

Moderator: We have time for one more question.

Col. Kovacs: Yes, please.

Audience Member 1: You mentioned that a lot of your students have an easier time dealing with cyber security and cyber warfare than electronic warfare. I was wondering if you could elaborate on that and perhaps speculate as to why that is and how that issue could be resolved.

Col. Kovacs: The main problem was the terminology. In the cyber security views, common terminology, because our students or our cadets in the beginning of the 20s, they are born when the computer was common, but electronic warfare came from the '60s. We use such a kind of phrase that electronic support measures. Nobody understands it. Electronic countermeasures. Nobody understands it. We have to explain what the hell the electronic support measure is. It's nothing new because this is the electronic intelligence in a special way, but today the cadets have no ideas. That's why it was a very, very important feedback for us. How can we, or it's compulsory to change the learning materials for them. We have to focus more, much more on explaining materials or on the explaining texts because of the terminology or the taxonomy. That's the main problem, or that was the main problem. It's not the students' problem. It's our problem, because we have been dealing with electronic warfare for many years. We know everything or almost everything mainly about the terminology, but we didn't think that it could be a new thing for them. Okay? Thank you.

Presentation Slides

NEMZETI KÖZSZOLGÁLATI EGYETEM
A HAZA SZOLGÁLATÁBAN

Cybersecurity as a Horizontal Issue in Public Service

Dr. Csaba Krasznay
Cybersecurity Academy
National University of Public Service, Hungary

NEMZETI KÖZSZOLGÁLATI EGYETEM
A HAZA SZOLGÁLATÁBAN

Motto

„This Strategy indicates that Hungary is ready to perform and take responsibility for cyberspace protection tasks and intends to develop the Hungarian cyberspace as a key element of Hungarian economic and social life into a free, secure and innovative environment. "

Government Decision No. 1139/2013 (21 March) on the National Cyber Security Strategy of Hungary

NEMZETI KÖZSZOLGÁLATI EGYETEM
A HAZA SZOLGÁLATÁBAN

Introduction

- „We can find the main source of information security problems between the chair and the keyboard."
- Therefore the Hungarian Cybersecurity Act emphasizes the importance of trainings.
- Under the cybersecurity framework the following actors should be trained
  - managers,
  - CISOs,
  - experts
  - supporters (e.g. IT operators),
  - employees,
  - The whole society.
- Cybersecurity is not solely an engineering issue anymore!
- Just think about the serious lack of human resources.

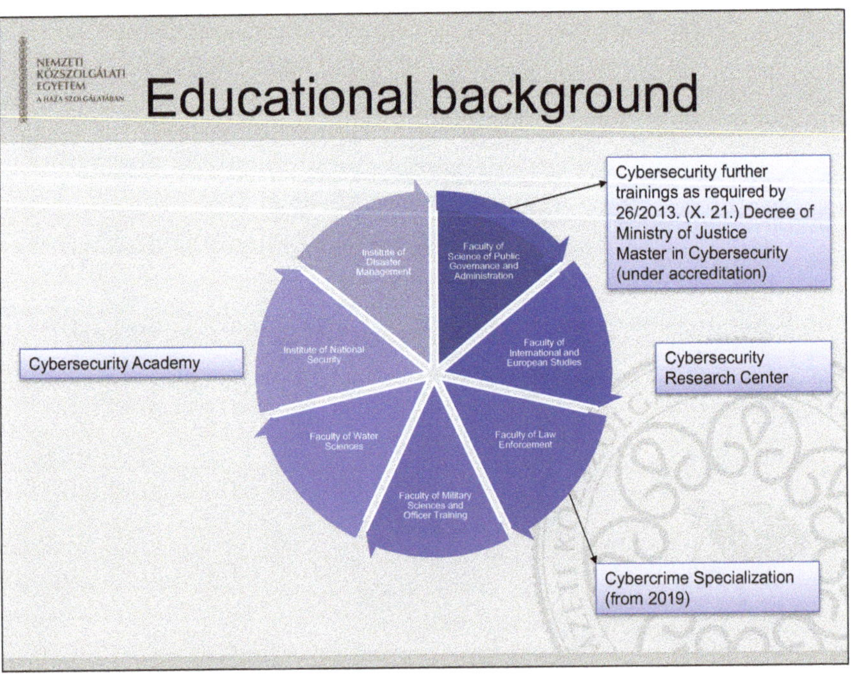

NEMZETI KÖZSZOLGÁLATI EGYETEM
Talent Management

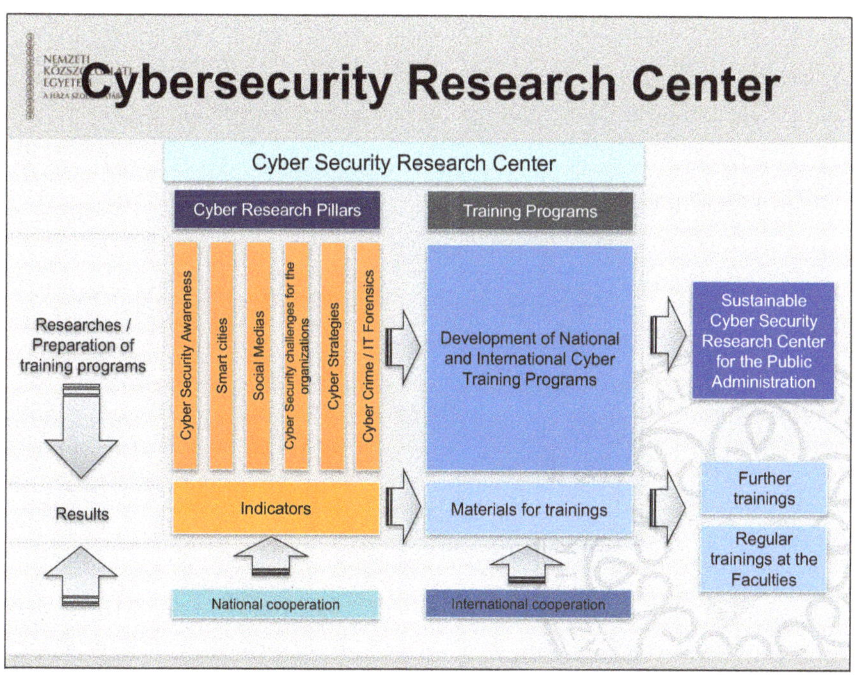
Cybersecurity Research Center
Cyber Security Research Center
Cyber Research Pillars
Training Programs
Cyber Security Awareness
Smart cities
Social Medias
Cyber Security challenges for the organizations
Cyber Strategies
Cyber Crime / IT Forensics
Researches / Preparation of training programs
Development of National and International Cyber Training Programs
Sustainable Cyber Security Research Center for the Public Administration
Results
Indicators
Materials for trainings
Further trainings
Regular trainings at the Faculties
National cooperation
International cooperation

NEMZETI KÖZSZOLGÁLATI EGYETEM A HAZA SZOLGÁLATÁBAN

NUPS Cybersecurity Academy

- Founded in March 2017
- It's main tasks are:
  - To synchronize all cybersecurity related educational and research activities
  - To cooperate with relevant stakeholders through the steering committee
  - To represent the university in national and international partnerships
  - To initiate and organize trainings, events and publications
  - To organize exercises and set up a cybersecurity laboratory

NEMZETI KÖZSZOLGÁLATI EGYETEM A HAZA SZOLGÁLATÁBAN

Steering Committee

- As cybersecurity is a real horizontal issue, the following stakeholders are presented:
  - All faculties and institutes from the university
  - Prime Minister's Office
  - Ministry of Defence
  - Ministry of Foreign Affairs and Trade
  - Ministry of Interior
  - Ministry of Justice
  - National Directorate General for Disaster Management, Ministry of the Interior
  - Constitution Protection Office
  - Military National Security Service
  - Special Service for National Security
  - National Authority for Data Protection and Freedom of Information
  - Hungarian National Police Cybercrime Unit
  - The Cybersecurity Coordinator of Hungary

NEMZETI KÖZSZOLGÁLATI EGYETEM
A HAZA SZOLGÁLATÁBAN

Goals

- Our goals are
  - To be a main actor in Hungary's cybersecurity framework (on the fields of public administration, law enforcement and military)
  - To support instant training tasks
  - To educate fresh experts
  - To edcuate cybersecurity aware public servants
  - To improve the public service life model
  - To lay down a high quality R&D background

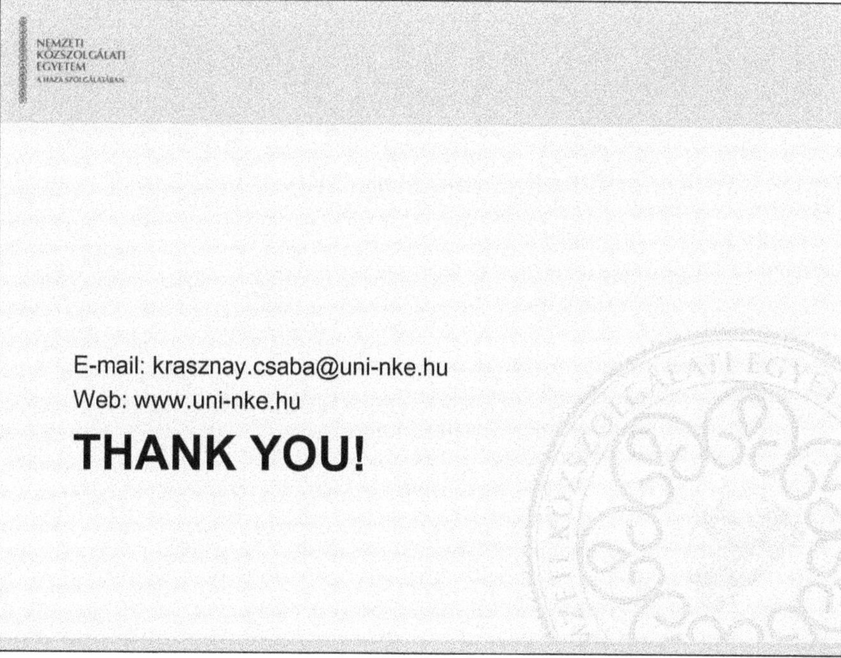

NEMZETI KÖZSZOLGÁLATI EGYETEM
E-mail: krasznay.csaba@uni-nke.hu
Web: www.uni-nke.hu
THANK YOU!

www.ingramcontent.com/pod-product-compliance
Lightning Source LLC
Chambersburg PA
CBHW042314210326
41599CB00038B/7126
9781940771519